ATP 3-05.20 (FM 3-05.120)

Special Operations Intelligence

May 2013

DISTRIBUTION RESTRICTION: Distribution authorized to U.S. Government agencies and their contractors only to protect technical or operational information from automatic dissemination under the International Exchange Program or by other means. This determination was made on 14 December 2012. Other requests for this document must be referred to Commander, United States Army John F. Kennedy Special Warfare Center and School, ATTN: AOJK-CDI-CID, 3004 Ardennes Street, Stop A, Fort Bragg, NC 28310-9610.

DESTRUCTION NOTICE: Destroy by any method that will prevent disclosure of contents or reconstruction of the document.

FOREIGN DISCLOSURE RESTRICTION (FD 6): This publication has been reviewed by the product developers in coordination with the United States Army John F. Kennedy Special Warfare Center and School foreign disclosure authority. This product is releasable to students from foreign countries on a case-by-case basis only.

Headquarters, Department of the Army

*ATP 3-05.20 (FM 3-05.120)

Army Techniques Publication
No. 3-05.20

Headquarters
Department of the Army
Washington, DC, 3 May 2013

Special Operations Intelligence

Contents

Page

PREFACE v

INTRODUCTION vi

Chapter 1 INTELLIGENCE AND ARMY SPECIAL OPERATIONS 1-1
Intelligence Assets 1-1
The Collaborative Information Environment 1-3
Intelligence Characteristics 1-3
Intelligence Process 1-5
Intelligence in the Military Decisionmaking Process 1-5
Characteristics of Effective Intelligence 1-6
Intelligence Capabilities and Disciplines 1-6
Intelligence Tasks 1-7
Intelligence Support to Personnel Recovery 1-9

Chapter 2 INTELLIGENCE SUPPORT TO TARGETING 2-1
Targeting and Planning 2-1
Joint Operational Design and Targeting Effects 2-1
Target Analysis Focus 2-2
Target Systems 2-8
Targeting 2-10

Distribution Restriction: Distribution authorized to U.S. Government agencies and their contractors only to protect technical or operational information from automatic dissemination under the International Exchange Program or by other means. This determination was made on 14 December 2012. Other requests for this document must be referred to Commander, United States Army John F. Kennedy Special Warfare Center and School, ATTN: AOJK-CDI-CID, 3004 Ardennes Street, Stop A, Fort Bragg, NC 28310-9610.

Destruction Notice: Destroy by any method that will prevent disclosure of contents or reconstruction of the document.

Foreign Disclosure Restriction (FD 6): This publication has been reviewed by the product developers in coordination with the United States Army John F. Kennedy Special Warfare Center and School foreign disclosure authority. This product is releasable to students from foreign countries on a case-by-case basis only.

*This publication supersedes FM 3-05.120, 15 July 2007.

Contents

	Combat Assessment	2-15
	Measures of Effectiveness	2-17
	Measures of Performance	2-17
Chapter 3	**ARMY SPECIAL OPERATIONS FORCES INTELLIGENCE SUPPORT SYSTEMS AND ARCHITECTURE**	**3-1**
	Connectivity, Systems, and Architecture	3-1
	Organic Intelligence Systems	3-2
	Army Special Operations Forces Databases and Tools	3-7
	Army Service Component Command Intelligence Support	3-9
	Joint and Theater-Level Intelligence Support	3-11
	Department of the Army Intelligence Support	3-12
	National-Level Intelligence Support	3-13
Chapter 4	**INTELLIGENCE SUPPORT TO CIVIL AFFAIRS**	**4-1**
	Mission	4-1
	Intelligence Requirements	4-1
	Intelligence Organization	4-2
	Nonorganic Intelligence Support	4-3
Chapter 5	**INTELLIGENCE SUPPORT TO MILITARY INFORMATION SUPPORT OPERATIONS**	**5-1**
	Mission	5-1
	Intelligence Requirements	5-1
	Intelligence Organization	5-3
	Nonorganic Intelligence Support	5-4
	Military Information Support Operations Support to the Intelligence Process	5-7
Chapter 6	**INTELLIGENCE SUPPORT TO RANGERS**	**6-1**
	Mission	6-1
	Intelligence Requirements	6-1
	Intelligence Organization	6-2
	Nonorganic Intelligence Support	6-7
	Ranger Support to the Intelligence Process	6-8
Chapter 7	**INTELLIGENCE SUPPORT TO SPECIAL FORCES**	**7-1**
	Mission	7-1
	Intelligence Requirements	7-1
	Intelligence Organization	7-1
	Nonorganic Intelligence Support	7-17
	Special Forces Support to the Intelligence Process	7-20
Chapter 8	**INTELLIGENCE SUPPORT TO SPECIAL OPERATIONS AVIATION**	**8-1**
	Mission	8-1
	Intelligence Requirements	8-1
	Intelligence Organization	8-2
	Nonorganic Intelligence Support	8-5
	Special Operations Aviation Regiment Support to the Intelligence Process	8-6
Appendix A	**OPEN-SOURCE INTELLIGENCE AND INFORMATION**	**A-1**
Appendix B	**TARGET ANALYSIS OUTLINE**	**B-1**

Contents

Appendix C	INTERAGENCY AND MULTINATIONAL INTELLIGENCE	C-1
Appendix D	LINGUIST SUPPORT	D-1
Appendix E	DOCUMENT EXPLOITATION AND HANDLING	E-1
	GLOSSARY	Glossary-1
	REFERENCES	References-1
	INDEX	Index-1

Figures

Figure 1-1. The military decisionmaking process 1-6
Figure 2-1. Joint targeting cycle 2-2
Figure 2-2. Target analysis focus 2-3
Figure 2-3. Time-sensitive targeting 2-14
Figure 2-4. F3EA targeting methodology 2-15
Figure 3-1. INTELINK 3-9
Figure 5-1. Potential intelligence sources for Military Information Support Operations 5-6
Figure 6-1. Regimental S-2 6-2
Figure 6-2. Ranger Military Intelligence Company organization 6-3
Figure 6-3. All-source analysis section 6-4
Figure 6-4. Single-source section 6-5
Figure 6-5. Ranger reconnaissance company organization 6-6
Figure 6-6. Ranger battalion S-2 organization 6-7
Figure 7-1. Special Forces group S-2 section 7-2
Figure 7-2. Special Forces group military intelligence detachment 7-4
Figure 7-3. Special operations task force operations center S-2 section 7-7
Figure 7-4. Special operations task force operations center military intelligence detachment 7-9
Figure 7-5. Battalion S-2 section 7-13
Figure 7-6. Cross Match biometrics kit 7-16
Figure 8-1. Intelligence organization of the Special Operations Aviation Regiment 8-2
Figure 8-2. Intelligence organization of the Special Operations Aviation Regiment battalion 8-3
Figure A-1. Internet components and elements A-3
Figure B-1. Sample target analysis outline B-1
Figure E-1. Example of a captured document log E-4

Tables

Table 1-1. Intelligence assets by organizations and echelon 1-2
Table 2-1. CARVER rating criteria 2-7
Table 2-2. CARVER value rating scale (notional) 2-10
Table 2-3. Sample strategic CARVER matrix application 2-11
Table 2-4. Sample operational CARVER matrix application 2-11
Table 2-5. Sample tactical CARVER matrix application 2-12
Table A-1. Invisible Web databases A-7
Table A-2. Military-accessible fee-based databases A-7
Table A-3. Source reliability A-8
Table A-4. Open-source information credibility A-8
Table A-5. Internet search techniques and procedures A-9
Table A-6. Boolean and math logic operators A-11
Table A-7. Internet domains A-13
Table C-1. Interagency and multinational organizations at the strategic, operational, and tactical levels C-2

Preface

Army Techniques Publication (ATP) 3-05.20, *Special Operations Intelligence*, provides the United States (U.S.) Army special operations forces (ARSOF) commander and his staff a broad understanding of intelligence support to, and the capabilities of, select ARSOF units to collect information and actionable intelligence. This publication also provides guidance for commanders who determine the force structure, budget, training, materiel, and operational requirements needed to prepare organic military intelligence assets to conduct their missions.

The principal audience of ATP 3-05.20 is ARSOF commanders and their staffs, supporting intelligence organizations, and supported joint and Army commands. ATP 3-05.20 is the unclassified techniques publication for intelligence support to ARSOF and supersedes Field Manual (FM) 3-05.120, *(S/NF) Army Special Operations Forces Intelligence (U)*. The acronym "ARSOF" represents Special Forces (SF), Rangers, special operations aviation (SOA), Military Information Support Operations (MISO), and Civil Affairs (CA). ATP 3-05.20 is an ARSOF subordinate doctrine publication and expands upon the doctrine in Army Doctrine Publication (ADP) 3-05, *Special Operations*. It should not be used independently, but rather, in conjunction with ADP 3-05 and Army Doctrine Reference Publication (ADRP) 3-05, *Special Operations*. It is authoritative doctrine for use by Regular Army and United States Army Reserve (USAR) ARSOF units and any other units or personnel operating with or supporting them.

This ARSOF doctrine is consistent with joint and Army doctrine. For a thorough understanding of intelligence support within the joint and land component environment, readers must also reference the following joint publications (JPs) and ADP/ADRP:

- JP 2-0, *Joint Intelligence*.
- JP 2-01, *Joint and National Intelligence Support to Military Operations*.
- JP 2-01.3, *Joint Intelligence Preparation of the Operational Environment*.
- JP 2-03, *Geospatial Intelligence Support to Joint Operations*.
- ADP 2-0, *Intelligence*.
- ADRP 2-0, *Intelligence*.

Commanders, staffs, and subordinates ensure their decisions and actions comply with applicable U.S., international, and, in some cases, host-nation laws and regulations. Commanders at all levels ensure their Soldiers operate in accordance with the law of war and the rules of engagement. (See FM 27-10, *The Law of Land Warfare*.)

ATP 3-05.20 uses joint terms where applicable. Selected joint and Army terms and definitions appear in both the glossary and the text. For definitions shown in the text, the term is italicized and the number of the proponent publication follows the definition. This publication is not the proponent for any Army terms.

ATP 3-05.20 applies to the Active Army, the Army National Guard/Army National Guard of the United States, and the USAR unless otherwise stated.

The proponent of ATP 3-05.20 is the United States Army John F. Kennedy Special Warfare Center and School (USAJFKSWCS). Send comments and recommendations on a DA Form 2028 (Recommended Changes to Publications and Blank Forms) to Commander, USAJFKSWCS, ATTN: AOJK-CDI-CID, 3004 Ardennes Street, Stop A, Fort Bragg, North Carolina 28310-9610; by e-mail to JACOMMENTS@soc.mil; or submit an electronic DA Form 2028.

Introduction

ATP 3-05.20 provides ARSOF commanders and their staffs, as well as supporting and supported agencies, organizations, and commands, techniques for providing intelligence support to U.S. ARSOF. In keeping this publication unclassified, it is more readily integrated with information provided by other doctrinal publications such as ADP/ADRP 3-05, *Special Operations*, and ADP/ADRP/FM 2-0, *Intelligence*, as well as doctrinal publications covering Civil Affairs, Military Information Support, Ranger, Special Forces, and special operations aviation units and their associated activities.

As stated in the preface, this publication supersedes FM 3-05.120, *(S/NF) Army Special Operations Forces Intelligence (U)*. In doing so, it omits classified information that pertained to the organization and capabilities of special operations units. Readers should not be concerned that classified techniques employed by special operations forces have been lost in the process. Those techniques were not included in the classified and now superseded field manual. Techniques and procedures that are classified are located in the appropriate ARSOF branch doctrinal publications. The change from classified to unclassified publication represents the only significant change in content. Other changes simply reflect updates in authoritative doctrine that are at a higher echelon in the Doctrine 2015 hierarchy, such as ADP/ADRP 3-05, *Special Operations*, and ADP/ADRP/FM 2-0, *Intelligence*.

This publication describes the missions and functions of intelligence elements and organizations that provide intelligence support to Army special operations units. It provides an overview of ARSOF intelligence capabilities, including the organization, missions, and functions of military intelligence personnel assigned to Army special operations units. It describes the connectivity from the level of the battalion intelligence staff to national-level intelligence organizations as ARSOF units execute missions within the range of military operations from peacetime engagement to major combat operations. It shows how organic and nonorganic assets meet operational needs within the intelligence process in order to provide relevant, accurate, predictive, and timely intelligence and information that allow special operations units to—

- Identify and develop targets.
- Develop and assess measures of effectiveness.
- Plan missions.
- Secure the element of surprise.
- Protect the force.

In addition, this publication describes the processes that unit commanders, staffs, and individual Soldiers use to collect both information and intelligence.

Chapter 1
Intelligence and Army Special Operations

U.S. Army special operations are often conducted with a high degree of risk to obtain high returns when national-level interests are at stake. Success for special operations dictates that uncertainties associated with the threat and environment are mitigated through the application of intelligence operations and procedures.

The commander executes the intelligence warfighting function throughout the range of military operations. The commander drives the content of the information collection plan by establishing his priority intelligence requirements and commander's critical information requirements. This chapter uses the doctrinal foundation in ADP/ADRP 2-0, *Intelligence*, and ADP/ADRP 3-05, *Special Operations*, to provide an overview of how the Army special operations units work with and within the intelligence enterprise.

INTELLIGENCE ASSETS

1-1. Special operations units' intelligence staffs must understand the intelligence requirements of their higher headquarters (HQ) and subordinate commands and components, and identify organic intelligence capabilities and shortfalls. The information and intelligence needs of Army special operations units frequently differ from their conventional counterparts in their scope and specificity, therefore requiring special operations unit intelligence staffs to exercise all seven intelligence disciplines. The joint nature of Army special operations and the need for detailed intelligence drive Army special operations unit intelligence staffs to use all the resources within the intelligence enterprise. By using all available resources they are able to overcome intelligence gaps and provide timely, relevant, accurate, and predictive intelligence.

1-2. The missions assigned to Army special operations units support the operational and strategic requirements of the geographic combatant commander (GCC). These missions are integrated with theater strategy and national objectives. Intelligence supports commanders with intelligence preparation of the battlefield (IPB), situation development, target development, support to targeting, information protection, fratricide avoidance, antiterrorism, survivability, safety, force health protection, and operational area security. Intelligence tasks include indications and warning and battle damage assessment. Intelligence organizations and assets that support Army special operations units are shown in Table 1-1, pages 1-2 and 1-3.

Table 1-1. Intelligence assets by organizations and echelon

Echelon	Producers	Organic Resources	Requests Support From
Theater	Combatant Command J-2 Joint Intelligence Center (JIC) Joint Intelligence Operations Center Special Operations Command J-2 Theater Army G-2	Joint Intelligence Support Element (JISE) Human Intelligence (HUMINT) Collectors Counterintelligence (CI) Agents Technical Intelligence (TECHINT) Collectors Signals Intelligence (SIGINT) Collectors Imagery Intelligence (IMINT) Collectors	Service Headquarters National Agencies Allies Army Commands
United States Special Operations Command (USSOCOM)	Special Operations Information Operations Staff Special Operations Command Joint Intelligence Center Joint Intelligence Operations Center	Analysts National Agency Liaisons	Theater JICs National Agencies Echelons Above Corps
United States Army Special Operations Command (USASOC)	G-2	Analysts Imagery Technicians CI Agents HUMINT Collectors National Geospatial-Intelligence Agency (NGA) Liaison Staff Weather Officer	USSOCOM Theater JICs Theater Special Operations Command (TSOC) J-2s
United States Army Special Forces Command (Airborne) (USASFC[A])	G-2	Analysts Imagery Technicians CI Agents HUMINT Collectors	USASOC Joint Task Force (JTF) J-2 Joint Special Operations Task Force (JSOTF) J-2
Special Forces Group (Airborne) (SFG[A])	Group S-2 Battalion S-2	Special Reconnaissance (SR) Teams Special Operations Team A (SOT-A) Analysts Imagery Technicians CI Agents HUMINT Collectors	USASFC(A) JTF J-2 JSOTF J-2
Ranger Regiment	Regiment S-2 Battalion S-2	Ranger Reconnaissance Company Analysts Imagery Technicians CI Agents HUMINT Collectors	USASOC JTF J-2 JSOTF J-2

Table 1-1. Intelligence assets by organizations and echelon (continued)

Echelon	Producers	Organic Resources	Requests Support From
Special Operations Aviation Regiment (SOAR)	Regiment S-2 Battalion S-2	Aircrews Analysts Imagery Technicians CI Agents HUMINT Collectors	USASOC JTF J-2 JSOTF J-2 Joint Special Operations Air Component Command J-2
CA Brigade	Brigade S-2 Battalion S-2	Analysts	USASOC JTF J-2 JSOTF J-2
Military Information Support Operations Command (Airborne) (MISOC[A])	G-2	Intelligence Analysts	USASOC JTF J-2 JSOTF J-2
Military Information Support Group (Airborne) (MISG[A])	Group S-2 Battalion S-2	Military Information Support (MIS) Detachment Cultural Intelligence Section Intelligence Analysts	USASOC JTF J-2 JSOTF J-2

NOTE: At all levels every Soldier is a sensor on the battlefield.

THE COLLABORATIVE INFORMATION ENVIRONMENT

1-3. The *information environment* is the aggregate of individuals, organizations, and systems that collect, process, disseminate, or act on information (JP 3-13, *Information Operations*). The collaborative information environment is a concept linking collaboration with infrastructure initiatives to produce not only ready paths of information exchange at the joint level, but the most current and accurate common operational picture possible. A collaborative information environment facilitates key collaborative functions to tie basic command processes to battlefield functions. Special operations forces (SOF) commanders and staffs use several informational and intelligence databases within the existing intelligence architecture to both transmit and use information and intelligence to develop what is increasingly a near-real-time common operational picture. Central to this is a focus on collaboration through a collaboration information environment.

INTELLIGENCE CHARACTERISTICS

1-4. Intelligence support to ARSOF is based on the same principles as intelligence support to conventional forces and uses the same intelligence process. However, intelligence requirements unique to Army special operations produce some distinct characteristics, as described below.

CRITICALITY

1-5. During special operations, especially during direct action or information-related activities conducted to influence targeted individuals and groups, there may be an opportunity to successfully execute a mission in denied areas. ARSOF commanders require extremely detailed and responsive intelligence support which is crucial in helping the commander make an informed decision during the military decisionmaking process (MDMP). Although all U.S. military forces need specific and detailed intelligence, a lack of sufficient detail on targets (in both lethal and nonlethal targeting options) may constrain or prohibit the use of Army special operations units to engage those targets.

Chapter 1

INTELLIGENCE CAPABILITIES

1-6. Assets and products derived from the unique mission capabilities of Army special operations units are valuable tools for combatant commanders. CI and HUMINT assets and products provide information or intelligence that may not be obtainable through other means. Army special operations units can provide alternate ways to acquire SIGINT and electronic warfare assets for combatant commanders to provide a more robust passive and active means to target adversaries and to protect the friendly mission command warfighting function and supporting systems. The cultural expertise of CA, MISO, and SF units provides a frequently unmatched capability to gain information considerations regarding the civil component of the operational environment and facilitate achievement of the commander's civil-military operations (CMO) desired end states.

INTELLIGENCE LIMITATIONS

1-7. ARSOF intelligence staffs above group or regiment level are limited in comparison with that of conventional operating forces. To overcome this limitation, commanders and their staffs must understand how the intelligence enterprise works and how to leverage intelligence assets and products to support their operations. The environments and types of operations will vary. Effective intelligence support to Army special operations units takes advantage of the intelligence enterprise's full capability.

NECESSITY FOR HUMAN INTELLIGENCE

1-8. Often accurate and relevant target-specific information critical for mission success can be obtained only through CI and HUMINT operations. CI and HUMINT information responds to reconnaissance and surveillance taskings, and supports information protection, fratricide avoidance, antiterrorism, survivability, safety, force health protection, operational area security, MISO series development and effectiveness assessment, CA, information operations coordinating and planning activities, target development, and the other intelligence disciplines. CI and HUMINT capabilities include host nation (HN) liaison, interrogation operations, screening operations, and CI force protection source operations. The intelligence collected from these activities could include plans of the internal layout of facilities, activities and intentions of security and threat forces, or potential areas for infiltration of assault forces. National-level technical collection means may be incapable of obtaining this specific level of data or lack the "loiter" time on the target needed to acquire information.

1-9. ARSOF often prefer speaking with those who have direct or firsthand knowledge and frequently favor information from HUMINT operations. This type of knowledge, once vetted and corroborated, often provides direct answers to information requirements. Examples include debriefing of U.S. and multinational personnel after they conduct operations and coming into contact with persons of interest. This preference is especially apparent when the HUMINT information is collected by special operations Soldiers. Examples include receiving information from a Special Forces operational detachment A (SFODA) SR mission and a Ranger Reconnaissance Company providing prestrike information to the commander of a special operations task force (SOTF).

USE OF OPEN SOURCES

1-10. The use of open-source intelligence (OSINT) is increasing at all echelons. Traditionally, Army special operations units have derived much of their preliminary operational intelligence needs through open-source information gathering. MISO and CA Soldiers have critical requirements for infrastructure, demographic, cultural, and psychological information on the areas and people on whom and through whom they operate. Use of tools such as direct observation, interviews, and exploiting information on the Internet are mainstays of MISO and CA information gathering. Credible studies and surveys conducted by intergovernmental and nongovernmental organizations often provide infrastructure, demographic, environmental hazard, socioeconomic, polling, and other data of value. The distinction between active and passive gathering is stressed to commanders; CA and MISO Soldiers only gather information passively. When relevant populations perceive CA and MISO personnel as strictly intelligence gatherers, the potential for losing credibility increases. Appendix A provides more information on OSINT.

Emphasis on Surprise, Security, and Deception

1-11. SOF are particularly sensitive to threat collection and targeting efforts. CI, operations security (OPSEC), and deception must be applied to protect sensitive missions across the range of military operations. Commanders at all levels should be well informed on the capability and effectiveness of threat intelligence and security services against ARSOF. The losses of security and/or of surprise are often abort criteria for special operations units.

1-12. Where deception is used, the intelligence staff must determine which threat sensors are available to collect the deception story, how much data to feed into the system to ensure the threat's intelligence agencies arrive at the desired conclusion, and how the threat decisionmaker will react to the deception effort. Intelligence and operations staffs collectively decide if organic special operations units and personnel will be used to support the deception (for example, MISO units targeting a false location and target audience with surrender leaflets) or if theater or national-level assets must be used to support or conduct the deception. The clandestine nature of some special operations activities demands exhaustive post-mission security measures to protect missions and participants. Security measures may require special operations personnel to interface with high-level elements of the Department of Defense (DOD) and other agencies.

Integrating Intelligence With Operations

1-13. The effective integration of intelligence with both current and future operations, as well as the timely and thorough debrief of operational elements, has long been a hallmark of battlefield success. Currently, commanders are adopting emerging techniques and procedures to further integrate their intelligence and operations staffs. These efforts have produced several different models at many levels of command. Tailored elements and working groups are emerging as an integrated operations and intelligence staff element or, in some cases, a new staff section concept. An example is the merging of the intelligence and operations staffs into a single staff section, referred to as the G-2/3 or S-2/3. Another example is the operations/intelligence fusion cell, also called the intelligence fusion cell. These may be small working groups with representatives from intelligence and operations staffs, or they may include other sections and representatives (for example, national agency liaisons). These cells may have only a few specific tasks, such as targeting and debriefing, or their task may be to synergize and deconflict operations and intelligence. This emerging concept has been used in military and homeland security operations with intelligence fusion cells including civilian law enforcement agencies and other government agencies.

INTELLIGENCE PROCESS

1-14. The intelligence process is designed to meet the commander's intelligence needs across the range of military operations. As discussed in ADP/ADRP 3-05, special operations are inherently joint, driving the requirement for intelligence personnel supporting special operations to be able to operate in both the Army and joint intelligence processes. The joint intelligence process consists of six interrelated categories of intelligence operations while the Army intelligence process consists of four steps and two continuing activities. In both processes, the commander's guidance drives the intelligence process. ADP 2-0 provides an overview of how the Army intelligence process accounts for the joint intelligence process categories. ADRP 2-0 provides details on the Army intelligence process and how it is synchronized with the operations process.

INTELLIGENCE IN THE MILITARY DECISIONMAKING PROCESS

1-15. The commander continually faces situations involving uncertainties, questionable or incomplete data, or several possible alternatives. As the primary decisionmaker, the commander, with the assistance of his staff, not only must decide what to do and how to do it (developing a final course of action [COA]), but he also must recognize if and when he must make a decision. The commander drives the MDMP, integrating his activities with those of his staff, subordinate HQ, and unified action partners. Effective intelligence supports this process contributing to the required application of sound judgment, logic, and professional knowledge.

Chapter 1

1-16. The MDMP (Figure 1-1) begins with the receipt of a mission by the commander. During the MDMP, the IPB process intensifies. The staff begins the evaluation of the area of operations, including clarifying the operational area with the operations staff and determining the area of interest. Special operations intelligence staffs are presented with the challenge of providing a detailed evaluation of the area of interest equal to the evaluation conducted for the operational area. Developing this level of detail for an area of interest is challenging for several reasons, including the geographic size of the area of interest. For instance, in a given operation, the MIS task force area of interest may be regional, transregional, or global.

Step 1: Receipt of mission.
Step 2: Mission analysis.
Step 3: COA development.
Step 4: COA analysis (war game).
Step 5: COA comparison.
Step 6: COA approval.
Step 7: Orders production, dissemination, and transition.

Figure 1-1. The military decisionmaking process

1-17. The intelligence staff begins by pulling from available intelligence databases, both organic and nonorganic. The intelligence staff performs terrain, climate, and areas, structures, capabilities, organizations, people, and events (ASCOPE) analysis, and then contacts the supporting special operations weather team (SOWT) for target weather information. The intelligence staff also analyzes the threat, determines its capabilities and vulnerabilities, prepares a situation template, and hypothesizes likely threat COAs. This basic process is applicable to any mission assigned to ARSOF.

1-18. The intelligence staff is responsible for production of the intelligence estimate. The intelligence estimate is the appraisal, expressed in writing or orally, of available intelligence relating to a specific situation or condition with a view to determining the COAs open to the threat and the order of probability of their adoption. This estimate consists of an analysis of the threat situation within the area of interest and the characteristics of the operational area in terms of how they can affect the mission. The intelligence staff uses the intelligence estimate to present conclusions and make recommendations, as appropriate. ADRP 2-0 provides further information on the intelligence estimate, including purpose, format, and content.

1-19. The MDMP provides the most thorough mission analysis approach available in time-constrained environments. The process leads to a solution to a problem (the "how" of conducting the mission) by analyzing in detail a number of friendly options against the full range of reasonable and available threat options. The resulting plan then serves as an optimum start point for later quick and effective adjustments as the unit begins its mission.

CHARACTERISTICS OF EFFECTIVE INTELLIGENCE

1-20. The effectiveness of intelligence is measured against two specific sets of criteria. The first measures the effectiveness against the information quality criteria of accuracy, timeliness, usability, completeness, precision, and reliability. The second set of characteristics is relevant, predictive, and tailored. ADRP 2-0 includes more information on these characteristics. When using these characteristics as criteria to measure effectiveness, it is important for the intelligence staff to be very specific in determining what the commander's requirements are relative to each criterion and recognize that these requirements may vary from mission to mission. While a certain degree of accuracy may be acceptable for some missions, it may not be high enough to be acceptable for others.

INTELLIGENCE CAPABILITIES AND DISCIPLINES

1-21. The intelligence enterprise is commonly organized through the intelligence disciplines. The joint function and the Army warfighting function of intelligence both employ intelligence capabilities to execute the intelligence process. The Army segregates its intelligence capabilities into all-source intelligence and single-source intelligence. The intelligence sources the capabilities refer to are the seven intelligence disciplines. The intelligence disciplines each have unique aspects of support that make organizing the intelligence enterprise by discipline almost natural. The challenge for Army special operations intelligence staffs is to integrate information from multiple disciplines. The seven intelligence disciplines are—

- Counterintelligence (CI).
- Geospatial intelligence (GEOINT).
- Human intelligence (HUMINT).
- Measurement and signature intelligence (MASINT).
- Open-source intelligence (OSINT).
- Signals intelligence (SIGINT).
- Technical intelligence (TECHINT).

1-22. There are a variety of complementary capabilities that affect intelligence operations. These complementary capabilities are not available Armywide and are unit/operation/mission specific. Biometrics, the Distributed Common Ground System-Army, document and media exploitation, and red teaming are a few examples of complementary capabilities. Technology drives many capabilities and continuing advances in technology have the Army working hard to keep doctrine up-to-date to support these capabilities. ADRP 2-0 provides more information on the Army intelligence capabilities and the intelligence disciplines.

INTELLIGENCE TASKS

1-23. The tasks performed by Army special operations unit intelligence staffs are not unique to special operations units. However, the focus and the scope of common joint and Army intelligence tasks conducted by special operations intelligence staffs often differ from those conducted by intelligence staffs that support a conventional unit. The following paragraphs describe areas of special emphasis and the unique focus of common intelligence tasks conducted by Army special operations staffs.

INTELLIGENCE PREPARATION OF THE BATTLEFIELD/BATTLESPACE

1-24. *Intelligence preparation of the battlefield/battlespace (IPB)* is a systematic process of analyzing and visualizing the portions of the mission variables of threat/adversary, terrain, weather, and civil considerations in a specific area of interest and for a specific mission. By applying intelligence preparation of the battlefield/battlespace, commanders gain the information necessary to selectively apply and maximize operational effectiveness at critical points in time and space (FM 2-01.3, *Intelligence Preparation of the Battlefield/Battlespace*). It is linked to but distinct from joint intelligence preparation of the operational environment (JIPOE). JIPOE and IPB products generally differ in terms of their relative purpose, focus, and level of detail. *Joint intelligence preparation of the operational environment* is the analytical process used by joint intelligence organizations to produce intelligence estimates and other intelligence products in support of the joint force commander's decision-making process. It is a continuous process that includes defining the operational environment; describing the impact of the operational environment; evaluating the adversary; and determining adversary courses of action (JP 2-01.3).

1-25. Army special operations unit intelligence staffs perform IPB in both joint and Army environments. The staff analyzes adversary and threat tactics, techniques, and procedures (TTP) using the mission variables already mentioned and operational variables—political, military, economic, social, information, infrastructure, physical environment, and time (PMESII-PT). This analysis develops required decision templates that depict high-value target patterns of activity and known networks of sustainment and support. Branches and sequels to the operation are established to help the commander and staff to develop a better understanding of the operational environment and identify potential decisions to be made because of ongoing operations. FM 2-01.3 provides more information on the IPB process.

Chapter 1

SITUATION DEVELOPMENT

1-26. Situation development is the process of continuously updating estimates of the situation and projections of threat capabilities and intentions. These estimates and projections let commanders see and understand the operational environment in enough time and detail to employ their forces effectively. Intelligence supports situation development by identifying intelligence requirements, developing a collection plan, monitoring indications and warnings problem sets, analyzing adversary activity, and providing intelligence assessments of adversary capabilities, vulnerabilities, centers of gravity, intentions, and possible COAs. Situation development incorporates all six operations of the intelligence process.

TARGET DEVELOPMENT AND SUPPORT TO TARGETING

1-27. Intelligence staffs know that the targeting process depends on effective and timely use of the intelligence process. Special operations target development is the result of complete and accurate situation development in all military operations. IPB supports target development and provides the commander with the intelligence needed to select valid targets.

BATTLE DAMAGE ASSESSMENT

1-28. Battle damage assessment is an analytic method to determine the relative success of a mission against an enemy target. Battle damage assessment is used to confirm or deny changes to enemy methods or patterns of activity that directly affect friendly operations. This makes battle damage assessment a predictive tool, as well as a measurement tool, for evaluating the degree of success of an operation, engagement, or battle. Battle damage assessment gives the commander a continual assessment of enemy strength and the effect of friendly operations on the enemy. It is a means to measure progress, to determine if another strike is needed, and to determine how close the unit is to accomplishing targeting goals.

INDICATIONS AND WARNING

1-29. Indications and warning is a critical subtask of the processing step of the intelligence process. This subtask is where the national strategic community monitors adversary/threat activity to determine if their diplomatic, informational, military, and economic actions are a prelude to hostilities or other acts contrary to U.S. interests. ARSOF are consumers and producers of indications and warning reports. They use indications and warning reports on world political-military developments to focus and refine intelligence databases and to update and guide contingency planning. Operational indications and warning intelligence becomes increasingly critical once a unit enters its final mission preparation and execution stages. Once deployed, special operations units can provide unique initial indications and warning reports from denied areas and can confirm or deny indications and warning reports from other sources.

ANALYZING CIVIL CONSIDERATIONS

1-30. In the course of conducting operations at a joint command or while in support of conventional Army units, Army special operations intelligence staffs may need to conduct civil considerations analysis in the context of how the standard ASCOPE analysis interfaces with operational and mission variable information. Army special operations intelligence Soldiers supporting joint commands may be required to conduct analysis using a modified ASCOPE analysis that includes synthesizing operational and mission variable information as it is related to ASCOPE. An operational and mission variable analysis approach integrates people and processes using multiple information sources and collaborative analysis to build a common, shared, and holistic knowledge base of the operational environment. Operational and mission variables analysis emphasizes a multidimensional approach toward situational understanding, distinguished by an analysis of the six interrelated characteristics of ASCOPE within each variable.

1-31. The application of the elements of ASCOPE during civil considerations analysis identifies the key and decisive areas, structures, capabilities, organizations, people, and events of each operational variable. A fully developed analysis will provide commanders and staff the ability to see the possible outcomes of missions conducted within the operational environment. It is through the analysis of the civil component that branches and sequels to the CMO plan are developed to assist the commander and staff in making critical decisions during operations in relationship to the civil component of the operational environment. FM 3-57, *Civil Affairs Operations*, provides more details on analyzing civil considerations.

INTELLIGENCE SUPPORT TO PERSONNEL RECOVERY

1-32. *Personnel recovery* is defined as the sum of military, diplomatic, and civil efforts to prepare for and execute the recovery and reintegration of isolated personnel (JP 3-50, *Personnel Recovery*). The following paragraphs discuss ARSOF intelligence support to personnel recovery.

PREDEPLOYMENT PERSONNEL RECOVERY INTELLIGENCE SUPPORT

1-33. During predeployment, intelligence staffs ensure that an isolated personnel report is on hand for all deploying Soldiers. Predeployment intelligence necessary to support Army special operations units also includes data that will support Soldiers in successfully rejoining their units should they become isolated personnel. The intelligence staff provides intelligence to support production of the commander's isolated personnel guidance. Isolated personnel guidance includes intelligence supporting survival, evasion, resistance, and escape. Intelligence products needed to produce the isolated personnel guidance include, but are not limited to—

- Escape and evasion route data.
- Local populace attitudes and hazards.
- Terrain, weather, and light data.
- Authentication and reporting requirements.

1-34. Intelligence staffs obtain the relevant joint personnel recovery support product for the operational area prior to deployment. The joint personnel recovery support product is the primary reference document for personnel recovery information on a particular country or region of interest. It is the successor to traditional printed intelligence products. Other intelligence products may be required to support preparation and initial-entry MISO seeking local populace support for personnel recovery (for example, blood chits, products specifying or implying rewards for support, and products specifying or implying retaliation for failure to support).

OPERATIONAL PERSONNEL RECOVERY INTELLIGENCE SUPPORT

1-35. Ongoing personnel recovery intelligence support focuses on obtaining survival, evasion, resistance, and escape and joint personnel recovery support product updates. Special operations unit intelligence detachments and staffs access and become familiar with the pertinent personnel recovery guidance or other published notification documents as they are updated regularly (for example, personnel recovery standing operating procedures, air tasking order special instructions, and isolated personnel guidance and survival, evasion, resistance, and escape updates). Updated intelligence on issues such as the treatment of enemy prisoners of war (EPWs) may be essential to ongoing MISO targeting detention facility personnel or enemy leadership as well as target audiences capable of assisting evaders.

NONORGANIC SUPPORT

1-36. Intelligence support for personnel recovery requires the production of highly detailed intelligence and typically involves national-level agencies and DOD intelligence organizations. Special operations units access this intelligence support through the same channels and mechanisms as other intelligence support. At the joint level, intelligence specific to personnel recovery is available through the joint personnel recovery center. JP 3-50 includes a detailed discussion of joint and national-level personnel recovery intelligence support.

UNCONVENTIONAL ASSISTED RECOVERY

1-37. Unconventional assisted recovery is a category of personnel recovery. Unconventional assisted recovery operations consist of unconventional warfare (UW) forces establishing and operating unconventional assisted recovery infrastructure and teams. Unconventional assisted recovery operations

Chapter 1

are designed to seek out, contact, authenticate, and support military and other selected personnel as they move from an adversary-held, hostile, or sensitive area to areas under friendly control. Personnel recovery missions may be executed by Army special operations units. It is more common that Army special operations units support personnel recovery operations conducted by dedicated joint assets and forces. SF units, in particular SFODAs conducting UW, plan and prepare for the contingency of conducting unconventional assisted recovery.

1-38. Army special operations units conducting unconventional assisted recovery operations require intelligence products that support the ongoing UW operation and products tailored for specific unconventional assisted recovery missions. No SFODA or other special operations unit supporting unconventional assisted recovery operations will be in possession of all the intelligence necessary to conduct an unconventional assisted recovery. Generally, detailed, specific intelligence products will need to be delivered to ARSOF units conducting unconventional assisted recovery, to include—

- Evader/escapee data, to include isolated personnel report data.
- Authentication data.
- IMINT of the isolated personnel's initial location and possible evasion routes.
- SIGINT on isolated personnel and/or adversary communications.
- MASINT of adversary activities (and isolated personnel, if applicable).
- Adversary capabilities.
- Weather effects.

Chapter 2
Intelligence Support to Targeting

Targeting is the process of selecting targets and matching the appropriate response and method of engagement. It is the analysis of enemy situations relative to the commander's mission, objectives, and capabilities at his disposal. It identifies and nominates specific vulnerabilities that, if exploited, will accomplish the commander's purpose through delaying, disrupting, disabling, or destroying enemy forces or denying resources critical to the enemy. The targeting process includes lethal and nonlethal vulnerabilities that can be exploited to accomplish the commander's operational end states.

TARGETING AND PLANNING

2-1. Deliberate evaluation of an adversary's vulnerabilities and application of SOF capabilities at critical nodes (to include individuals) are the foundation of special operations unit employment. SOF select targets for exploitation with careful and deliberate consideration. This is the hallmark of special operations planning and is best accomplished by the element that will execute the plan. Effective integration of special operations into the GCC's campaign is possible only through synchronized targeting and mission planning.

2-2. A joint force commander (JFC) may establish a joint targeting coordination board to evaluate nominations to assess whether targets will achieve desired objectives. The options available to the JFC include lethal and nonlethal effects that can optimize special operations units' capabilities. Special operations targeting considerations include the political, military, economic, and psychological impact on the enemy force's capabilities and morale and its popular support base. Two distinctly different modes—direct and indirect—define the lethal and nonlethal force applied by ARSOF.

2-3. The joint targeting cycle portrays an analytical, systematic approach focusing on the targeting process that supports operational planning to achieve the objectives of the JFC. The interrelationship of the target development and mission planning phases dominates the six phases of the joint targeting cycle (Figure 2-1, page 2-2). All of these phases require proactive and synchronized intelligence support. Intelligence staff officers supporting special operations units must anticipate mission needs by the development of branches and sequels to the operational plan and coordinate for the resources required to provide the timeliest, most relevant, accurate, and predictive intelligence possible.

JOINT OPERATIONAL DESIGN AND TARGETING EFFECTS

2-4. Considering the elements of operational design during targeting combines both lethal and nonlethal options to achieve a commander's desired effects in order to achieve strategic end states (JP 5-0, *Joint Operation Planning*). The targeting process is a vital link between tactical tasks and strategic end states. Targeting effects are the cumulative results of actions taken to engage geographical areas, complexes, installations, forces, equipment, functions, perception, or information by lethal and nonlethal means. Targeting effects are either direct or indirect. Direct effects are the immediate, first-order consequence of a military action. Indirect effects are the delayed and/or displaced second- and third-order consequences of military action. In addition, effects possess three fundamental characteristics that qualitatively impact the influence they exert on adversary capabilities:

- Cumulative nature of effects.
- Cascading nature of effects.
- Collateral and additional nature of effects.

Chapter 2

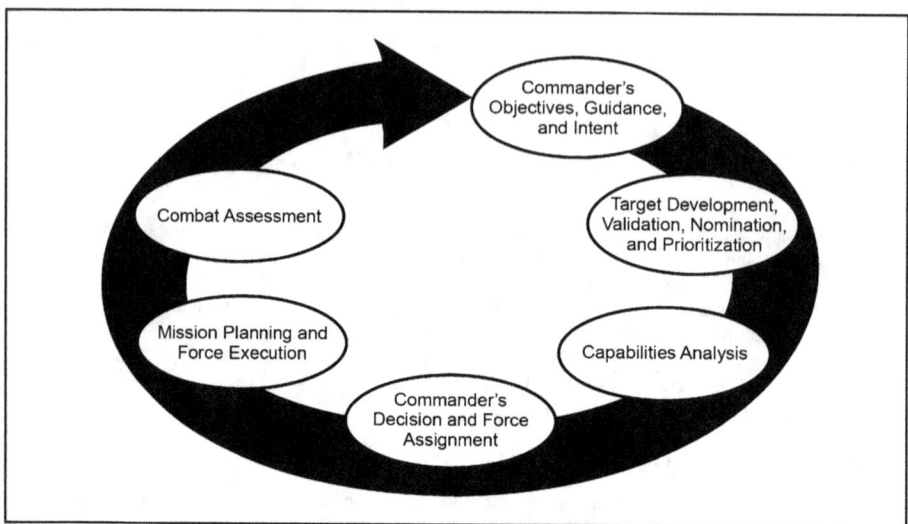

Figure 2-1. Joint targeting cycle

2-5. Army special operations units implement a planning and intelligence collection effort that considers the political, psychological, and long-term second- and third-order effects of their employment. This collection effort is applied as a result of operational design and may focus on specific elements of operational design. Sufficient intelligence to provide a predictive framework to identify and quantify second- and third-order effects of tactical tasks must be available through the targeting process to determine which capabilities will be brought to bear to produce a specific effect. This specific effect is usually one of many needed to achieve strategic end states.

TARGET ANALYSIS FOCUS

2-6. The target analysis methodology as outlined in JP 3-05.1, *Joint Special Operations Task Force Operations*, examines potential targets to determine their military importance, priority of attack scale of effort, and weapons required to attain a certain level of damage, disruption, or lethal or nonlethal casualties. It is a systematic approach to establishing the enemy vulnerabilities and weaknesses to be exploited. It also determines what effects can be achieved against target systems and their activities. A target analysis must review the subsystems and interactions between components and elements of a target system to determine how the overall system functions and, subsequently, how to best attack that system so that it becomes inoperable or allows for achievement of the commander's objectives. The challenge for military intelligence personnel is to apply the intelligence process to not only the immediate target being considered but to anticipate second- and third-order effects and develop predictive intelligence on them as well. The following sections will give an overview of center of gravity analysis (as a primary means of identifying priority targets), the target analysis methodology, and an overview and examples of the criticality, accessibility, recuperability, vulnerability, effect, and recognizability (CARVER) factors matrix.

2-7. TSOC planners focus on the strategic and operational levels, whereas most JSOTFs evaluate targets from an operational and tactical perspective. JSOTF components designated as mission planning agents focus on the tactical level (Figure 2-2, page 2-3). At the strategic level, the analyst determines which target system to attack. For example, the strategic analyst may determine that disabling the power system in a country would cause factories to close, thereby undermining economic stability. At the operational level, the analyst determines which subsystem to attack. For instance, the operational analyst may determine that disabling a particular power plant would lead to failure of the entire power system or grid in the country or

Intelligence Support to Targeting

joint operations area. At the tactical level, the analyst determines which component of a target or node should be attacked or influenced to achieve the desired effect. In this case, a tactical analyst may determine that disabling a particular turbine or boiler would result in disabling the targeted power plant. Similarly, an analyst may determine a nonlethal influence operation targeted at a tribal chief may produce the desired effects. A target may consist of a target system, a subsystem, a complex, or a component or critical point. Intelligence personnel must remember that for special operations missions this component may be a single-named individual or group.

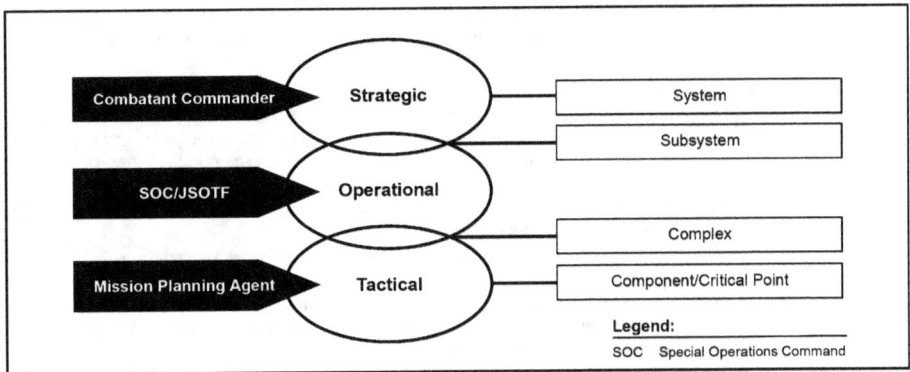

Figure 2-2. Target analysis focus

2-8. Normally, the TSOC staff conducts analysis at the strategic and operational levels during deliberate planning while determining how to support the GCC's strategic objectives. The JSOTF/SOTF conducts analysis at the operational level during crisis action planning execution phase planning when determining how to satisfy JTF objectives. They should also consider the tactical level of analysis, but it should not dictate to the mission planning agent how to affect the target. The JSOTF/SOTF should understand the intelligence and operational information requirements of the mission planning agent and facilitate quick, collaborative feasibility assessments and planning by requesting and expediting the required information to the mission planning agent. The mission planning agent conducts tactical-level analysis during deliberate or execution phase planning.

THREAT CENTER OF GRAVITY ANALYSIS

2-9. A *center of gravity* is defined as the source of power that provides moral or physical strength, freedom of action, or will to act (JP 5-0). At the tactical level, an objective is the equivalent to a center of gravity. At the strategic level, the center of gravity may be vulnerable to an operational-level approach. At the operational level, the center of gravity may be vulnerable to tactical actions. The purpose of performing a threat center of gravity analysis is to determine and evaluate critical vulnerabilities for exploitation. The results of center of gravity analysis are later used during COA development to exploit identified vulnerabilities.

2-10. During targeting, identified centers of gravities and objectives may be targeted with both lethal and nonlethal fires by Army special operations units. Essential definitions associated with center of gravity analysis are—

- *Critical capability*: A means that is considered a crucial enabler for a center of gravity to function as such and is essential to the accomplishment of the specified or assumed objective(s). (JP 5-0)
- *Critical requirement*: An essential condition, resource, and means for a critical capability to be fully operational. (JP 5-0)

- *Critical vulnerability*: An aspect of a critical requirement which is deficient or vulnerable to direct or indirect attack that will create decisive or significant effects. (JP 5-0)

2-11. Accurate intelligence assessments of adversary centers of gravity and critical vulnerabilities are essential to the targeting cycle. Special operations unit intelligence staffs support center of gravity analysis by providing multidiscipline intelligence on potential centers of gravity that can be engaged by Army special operations units. During a foreign internal defense operation, the top opposition leadership may be a center of gravity; during offensive operations, the communications infrastructure may be a center of gravity. Special operations units' intelligence staffs have the ability to provide center of gravity analysis throughout the range of military operations. Center of gravity analysis is conducted through the following steps:

- *Identify threat centers of gravity*. Visualize the threat as a system of functional components. Based upon the way the threat organizes, fights, and makes decisions, and upon its physical and psychological strengths and weaknesses, select the threat's primary source of moral or physical strength, power, and resistance. Centers of gravity may be tangible entities or intangible concepts. To test the validity of centers of gravity, ask the following question: Will the destruction, neutralization, or substantial weakening of the center of gravity result in changing the threat's COA or in denying its objectives?
- *Identify critical capabilities*. Each center of gravity is analyzed to determine what primary abilities (functions) the threat possesses in the context of the operational environment and friendly mission that can prevent friendly forces from accomplishing the mission. Critical capabilities are not tangible objects but rather are threat functions. To test the validity of critical capabilities, ask the following questions:
  - Is the identified critical capability a primary ability in context with the given missions of both threat and friendly forces?
  - Is the identified critical capability directly related to the center of gravity?
- *Identify critical requirements*. Each critical capability is analyzed to determine the conditions, resources, or means that enable threat functions or missions. Critical requirements are usually tangible elements, such as communications means, weapons systems, geographical areas, or terrain features. To test the validity of critical requirements, ask the following questions:
  - Will the absence or loss of the identified critical requirement disable the threat's critical capability?
  - Does the threat consider the identified critical requirement to be critical?
- *Identify critical vulnerabilities*. Each critical capability is analyzed to determine which critical requirements, or components thereof, are vulnerable to neutralization, interdiction, or attack. A critical vulnerability may be a tangible structure or equipment, or it may be an intangible perception, populace belief, or susceptibility. To test the validity of critical vulnerabilities, ask the following questions:
  - Will exploitation of the critical vulnerability disable the associated critical requirement?
  - Does the friendly force have the resources to impact the identified critical vulnerability?
- *Prioritize critical vulnerabilities through CARVER*.

2-12. At each step, intelligence supports center of gravity analysis by quantifying the concept of "criticality." Doing so consists of supporting the identification and quantifying both tangible (objects) and intangible (functions) elements of the operational environment. Analysts examining critical capabilities require a wide range of intelligence products from all disciplines. For instance, HUMINT plays a vital role in assessing the ability of an adversary force to maintain effective command in a highly centralized structure. When assessing that same adversary's ability to effectively communicate, the use of SIGINT, MASINT, and IMINT products is decisive. Critical requirements, with their focus on tangible systems, may not require as many HUMINT resources. Critical vulnerabilities may be HUMINT resource intensive in the case of analysis of enemy perceptions. In the case of structural or other physical targets, critical vulnerabilities may be GEOINT, SIGINT, and TECHINT resource intensive. Centers of gravity are invariably complex, and the preceding examples are not exhaustive; multiple sources and disciplines will be required.

2-13. The unique capabilities of ARSOF combined with the high value and high payoff of centers of gravity and their component parts prompt the tasking of special operations units to support engagement activities in addition to being the weapon of choice for engagement. Army special operations units may be tasked to conduct SR missions to identify centers of gravity. Special operations units may be tasked to assess measures of performance or measures of effectiveness of attacking elements and engaged targets to determine whether or not component missions, campaign phases, or a theater campaign are achieving the JFC's objectives.

TARGET ANALYSIS PROCESS AND TARGET SELECTION

2-14. Sound target analysis provides options to planners, satisfies a statement of requirement, meets the commander's objectives, and reduces risk to the operational element. Operations without good target analysis may waste effort and resources, introduce unnecessary risks, jeopardize future operations, and ultimately result in mission failure. Often, the operational element conducts target analysis without a specified statement of requirement. This situation may occur during deliberate peacetime planning in support of theater operation plans, during support of area studies, or during training to develop expertise in specific industrial systems located in a target country.

2-15. The commander's objectives are expressed in a statement of requirements, which is outlined in Sections II and III of the mission tasking package in the special operations mission planning folder. The statement of requirements defines the objectives and goals of analysis. It may be explicit or implied. Essential elements of a statement of requirements should answer the following questions:

- What is desired?
- What is the target or mission?
- What are the main and secondary targets?
- What are the targets of opportunity?
- Is the mission overt or clandestine?
- What extent of damage or downtime is desired?
- What is the strategic importance of the system or subsystem to be analyzed?
- What is the purpose of the mission?
- What is the deadline?
- When must analysis be completed?
- What are the force limitations?

The target analysis process involves the following steps and substeps:
- Step 1: Examine the statement of requirements.
    - Identify the target.
    - Identify the desired results.
    - Identify the operational constraints.
    - Identify the shortfalls.
    - Develop a collection plan.
    - Task the collectors.
- Step 2: Collect target intelligence, using—
    - CI.
    - GEOINT.
    - HUMINT.
    - MASINT.
    - OSINT.
    - SIGINT.
    - TECHINT.

Chapter 2

- Step 3: Collate the intelligence by systems, subsystems, complexes, and components.
  - Identify shortfalls.
  - Develop a collection plan.
  - Task the collectors.
- Step 4: Draw a flow chart that identifies—
  - Systems, subsystems, or complexes.
  - Critical nodes and choke points.
- Step 5: Evaluate systems, subsystems, or complexes by applying the factors of CARVER.
- Step 6: Identify potential targets.
  - Draft a preliminary list.
  - Refine the list using CARVER.
  - Eliminate the obvious.
- Step 7: Recommend appropriate targets that best satisfy the statement of requirements.
- Step 8: List the targets and their recommended priority.
  - Do they satisfy the statement of requirements?
  - Are they rational?
- Step 9: Obtain final target decision from commander.

2-16. When all the evaluation criteria have been established, the analyst uses a numerical rating system (1 to 5 or 1 to 10, depending on the unit standing operating procedure) to rank order the CARVER factors (Table 2-1, page 2-7) for each potential target. In a 1-to-5 numbering system, a score of 5 indicates a highly desirable rating (from the attacker's perspective), while a score of 1 reflects an undesirable rating. The analyst must tailor the criteria and rating scheme to suit the particular strategic, operational, or tactical situation. A sample target analysis outline is shown in Appendix B.

2-17. *Criticality*, or target value, is the importance of the target. It is the primary consideration in targeting. Criticality is related to how much a target's destruction, denial, disruption, and damage will impair the adversary's political, economic, or military operations or how much a target component will disrupt the function of a target complex. In determining criticality, individual targets within a target system must be analyzed with relation to the other elements critical to the function of the target system or complex. Critical targets may also be selected for SR missions.

2-18. *Accessibility* is the ease with which a target can be reached. To damage, destroy, disrupt, deny, or collect data on a target, personnel must be able to reach the target with the required system, directly or indirectly. During SR missions, personnel must not only observe the target but also remain in the area undetected for extended periods of time. The unit must also be able to exfiltrate safely once the mission is complete. Weather, light data, physical security measures, and the adversary's disposition at the target area are all considered. Sometimes, accessibility is judged as either feasible or infeasible.

2-19. *Recuperability* is a measure of the time required to replace, repair, or bypass the destruction or damage inflicted on a target. In direct action missions, an estimate must be made on the amount of time the adversary will take to repair, replace, or bypass the damage inflicted on a target. Primary considerations are the availability of spare parts and the ability to reroute production. Special operations units are not used to engage targets that can be repaired or bypassed in a short amount of time or with minimal resources.

2-20. *Vulnerability* is a measure of the ability of the action element to damage the target using available assets. A target is vulnerable if the special operations unit has the means and expertise to attack the target. At the strategic level, greater resources and technology are available to conduct the target attack. At the tactical level, resources may be limited to organic personnel, weapons, and munitions or to assets attached, borrowed, or improvised.

2-21. *Effect* is the positive or negative influence on the population as a result of the action taken. The target should be attacked only if the desired military effects can be achieved. These effects may be of a military, political, economic, informational, or psychological nature. The effect on the populace is viewed in terms of alienating the local inhabitants, strengthening the resistance movement, or triggering reprisals against the indigenous people in the immediate target area. The effect on the populace may impact infiltration, exfiltration, and evasion and recovery routes. Collateral damage must also be calculated and weighed against the expected military benefit to determine if an attack is advisable under the concept of proportionality. Collateral damage includes, but is not limited to, civilian injuries or deaths, damage to protected sites, and adverse economic impact of the proposed attack.

2-22. *Recognizability* is the degree to which a target can be identified under varying weather, light, and seasonal conditions without being confused with other targets or components. Sufficient data must be available for units to locate the target and distinguish it from similar objects in the target area. The same requirement exists to distinguish target critical damage points and target stress points from similar components and their parent structures and surroundings

Table 2-1. CARVER rating criteria

Criticality	Rating
Immediate output halt or 100 percent curtailment.	10
Output halt in less than 1 day or 75 percent curtailment.	6
Output halt in less than 1 week or 50 percent curtailment.	4
Output halt in over 1 week or less than 25 percent curtailment.	1
Accessibility	**Rating**
Standoff weapons can be employed.	10
Inside perimeter fence, but outdoors.	8
Inside building, but ground floor.	6
Inside building, but second floor.	4
Inside building, climbing required.	1
Recuperability	**Rating**
One month or more.	10
Up to 1 month.	8
Up to 1 week.	6
Up to 1 day.	4
Four hours or less.	1
Vulnerability	**Rating**
Small arms fire or charges of 5 pounds or less.	10
Antitank weapons or charges of 5 to 10 pounds.	7
Charges of 10 to 30 pounds.	5
Charges over 30 pounds or if laser-guided munitions must be deployed.	3
More drastic measures required.	1
Effect	**Rating**
National objectives fostered; no reprisals likely against friendly forces.	10
No effect or neutral.	5
Very negative public reaction; reprisals likely against friendly forces or high domestic negative reaction potential.	1
Recognizability	**Rating**
The complex or component is recognizable day or night, rain or shine, without confusion with other complexes or components.	10
The complex or component may be difficult to recognize at night or in bad weather or might be confused with other complexes or components.	5
The complex or component is difficult to recognize under any condition and is easily confused with other complexes or components.	1

Chapter 2

TARGET SYSTEMS

2-23. Both lethal and nonlethal methods need to be considered and available for employment to successfully complete missions through decisive action. For example, an assault may be launched against enemy forces, while simultaneously executing MISO against hostile political targets and engaging friendly or neutral indigenous civilian organizations with Civil Affairs operations (CAO). Planners involved in the targeting process must avoid merely focusing on the destructive or lethal effects based upon identifying enemy vulnerabilities and critical centers of gravity, and striking critical nodes to achieve a synergistic effect. Failure to analyze the operational variables (PMESII-PT) of the operational environment may cause inadequate responses to the underlying reasons of the instability and, in turn, may initiate or prolong a conflict. These complementary nonlethal effects contribute to this process and should not be ignored.

2-24. Stability operations are now recognized as being equal to combat operations, and targeting operations should reflect this reality. Special operations planners look at the applicable elements of operational design for the specific operation and identify supporting systems of the element. Increased emphasis in stability operations, especially in urban areas, stresses the need to analyze the effects of the socioeconomic systems—demographics, ethnic information, historical background, political and religious tensions and conflicts, anachronistic customs and behaviors, relative levels of corruption, suspicion of the government, criminal unrest, ascribed traditions and norms, and levels of political mobilization and polarization. In this context, special operations Soldiers conduct lethal and nonlethal target analysis for industrial and socioeconomic systems. These are the systems considered most often during irregular warfare and are of primary concern during all special operations before the introduction of conventional operating forces. The eight target systems, described in the following paragraphs, characterize the systems most prevalent in and critical to urban areas. Heavily populated areas receive considerable attention during UW operations as they provide opportunity to reach large numbers of people through a single engagement activity.

2-25. Virtually all target systems have critical subsystems. Typically, the successful targeting of any one of these subsystems results in the failure of the system itself. The interdiction of some subsystems affects the target system for longer periods than others. Accurate intelligence is vital in determining which subsystems will be targeted to achieve the commander's desired effect. Detailed collection and analysis of intelligence on subsystems is a critical enabler for effective targeting. The following paragraphs provide an overview of critical systems and their typical subsystems that special operations unit intelligence staffs routinely focus their efforts on.

ELECTRIC POWER

2-26. Abundant and dependable electric power supply is essential to modern industrial and military activities. The serious interruption or curtailment of bulk electric power supply can erode the warfighting capabilities of an adversary or weaken public resolve and support. For the purposes of target analysis, the subsystems of bulk electric power supply may be categorized as generation, transmission, control, and distribution.

PETROLEUM SUPPLIES

2-27. The petroleum supply industry is second only to the electric power supply industry in the relative order of importance in a national war effort. Petroleum, oil, and lubricants are essential to factory production, electric power generation, and virtually all forms of transport. For the purposes of analysis, the subsystems of bulk petroleum supply are production, long-distance transport and transmission, refining, and distribution.

WATER SUPPLIES

2-28. Water is the lifeblood of modern industrial and agricultural societies. Water is a key industrial target because it is essential for consumption, sanitation, fire fighting, and industrial and agricultural purposes. For target analysis, the subsystems of a large water supply system may be categorized as collection, transmission, treatment, and distribution.

COMMUNICATIONS

2-29. Redundant and dependable communications are essential to modern industrial and military activities. An effective communications system employs a synergistic architecture linking mission command nodes to supply a continuous automated flow and processing of information through rapid and secure voice, data, facsimile, and video communications. The serious disruption or degradation of major communications systems can drastically reduce an adversary's ability to wage war by restricting the rate and flow of vital information, thereby preventing the rapid and effective transmission of orders to influence the decisionmaking process. For the purposes of analysis, the subsystems of a communications system are local telephone networks, switching centers, control centers, signal transfer points, and transmission media.

AIR TRANSPORTATION

2-30. Air transportation is a vital military commodity before and at the outbreak of hostilities. The degradation of air transport capabilities can seriously affect a nation's ability to wage war. For the purposes of analysis, the major subsystems of an air transportation system are aircrews, aircraft, command, control, navigational aids, and airfield support services.

PORTS AND WATERWAYS

2-31. Large ports and waterway transportation systems are important to the economic, political, and military infrastructures of many countries. Serious disruption of port and waterway operations can undermine a nation's ability to conduct long-range military activities. Restricting the rate and flow of supporting war materiel by channeling the mobility of personnel and equipment increases pressure on other modes of transport and causes choke points that can present lucrative targets for aerial and other forms of interdiction. For the purposes of analysis, the major subsystems are vessels, waterways, ports, shipbuilding and repair facilities (dry dock), and services facilities.

RAILWAYS

2-32. Railroads represent an important transportation system in most countries. Railroads transport more freight than roads, waterways, and airports combined. Railroads are critical to a national economy, particularly in wartime, since they are required for the mass movement of personnel and materiel. For the purposes of analysis, the major subsystems are control facilities and signaling links; yards, terminals, and control centers; tracks and structures; and rolling stock.

SOCIOECONOMIC SYSTEM

2-33. Generally, the socioeconomic system has an immediate impact on the indigenous population in the operational area and influences larger, long-term political, economic, and informational areas that directly impact military operations. An appreciation of the socioeconomic system enhances the commander's selection of objectives in relation to the location, movement, and control of forces; the selection and use of weapons systems; and the protection measures. Second- and third-order effects may be the focus of analysis as disabling the socioeconomic system may target the will of those in that system to continue resistance.

2-34. For the purposes of analysis, the major subsystems of a socioeconomic system are the civil areas of key civilian interest, defined by political boundaries; structures, including protected targets; capabilities of public or government infrastructures; organizations at the local government or nongovernment level; people, military and civilian, as a target population; and events or significant activities that influence the civilian population. In MISO, target audiences consist of organizations, demographic sets, leaders, and key communicators.

RELATED CONSIDERATIONS

2-35. Although road networks are not a separate target system, they are closely related to many of the other target systems and should be considered during target analysis. Road networks are important avenues to the economic, political, social, and military infrastructure of any nation. The transportation of goods and services are critical for the overall economic viability of any country to wage warfare. The control or denial of critical avenues of approach or the supporting infrastructures reduces the effectiveness of all military operations. For the purposes of analysis, the major subsystems of road networks are bridges, tunnels, and major intersections.

TARGETING

2-36. The joint targeting process facilitates the publication of special operations tasking orders by providing amplifying information for detailed force-level planning of operations. Military operations and CMO occur simultaneously in all operational environments. ARSOF also focus on the various nonmilitary aspects of the operation as well as defeating a defined adversary. They engage the industrial component of the operational environment at the strategic, operational, and tactical levels. Eventually, military destructive engagements give way to civilian constructive engagements. The knowledge, experience, and perspective of special operations Soldiers facilitate this transition from the early planning stages to the final redeployment of U.S. forces. CARVER analysis focuses on the destructive effects of industrial target systems, and the socioeconomic system is engaged by CMO and MISO under nonlethal targeting.

CARVER EVALUATION CRITERIA

2-37. Once all the evaluation criteria have been established, the SOF analyst uses a numerical rating system (1 to 5) to rank order the CARVER factors for each potential target. In a 1-to-5 numbering system, a score of 5 would indicate a very desirable rating (from the attacker's perspective) while a score of 1 would reflect an undesirable rating. A notional CARVER value rating scale is shown in Table 2-2. The analyst must tailor the criteria and rating scheme to suit the particular strategic, operational, or tactical situation.

Table 2-2. CARVER value rating scale (notional)

Value	C	A	R	V	E	R	Value
5	Loss would be mission stopper.	Easily accessible. Away from security.	Extremely difficult to replace. Long downtime (1 year).	SOF definitely have the means and expertise to attack.	Favorable sociological impact. Okay impact on civilians.	Easily recognized by all with no confusion.	5
4	Loss would reduce mission performance considerably.	Easily accessible outside.	Difficult to replace with long downtime (< 1 year).	SOF probably have the means and expertise.	Favorable impact; no adverse impact on civilians.	Easily recognized by most, with little confusion.	4
3	Loss would reduce mission performance.	Accessible.	Can be replaced in a relatively short time (months).	SOF may have the means and expertise to attack.	Favorable impact; some adverse impact on civilians.	Recognized with some training.	3
2	Loss may reduce mission performances.	Difficult to gain access.	Easily replaced in a short time (weeks).	SOF probably have no impact.	No impact; adverse impact on civilians.	Hard to recognize; confusion probable.	2
1	Loss would not affect mission performance.	Very difficult to gain access.	Easily replaced in a short time (days).	SOF do not have much capability to attack.	Unfavorable impact; assured adverse impact on civilians.	Extremely difficult to recognize without extensive orientation.	1
NOTE: For specific targets, more precise target-related data can be developed for each element in the matrix.							

2-38. While the primary purpose of CARVER analysis is to quantify destructive effects of target systems, it is also a necessity for analysts to factor in second- and third-order effects of potential strikes on every target in order to avoid undesired effects. For example, disabling an enemy rail network by destroying it in its entirety may allow for the civilian unrest necessary to ultimately unseat a regime, but the long-term effects of that destroyed network would most likely hinder support to the government assuming power in the regime's wake. Disabling only a critical subcomponent of the rail network, however, would allow for a more timely, cost-effective restitution of civil infrastructure following cessation of hostilities.

STRATEGIC CARVER EVALUATION CRITERIA

2-39. The purpose of strategic target analysis is to determine the critical systems or subsystems that must be attacked for progressive destruction or degradation of an adversary's warfighting capacity and will to fight. Strategic operations have long-range, rather than immediate, effects on an adversary and its military forces. Strategic target analysis lists an adversary's systems or subsystems; for example, electric power and rail facilities. The results of the strategic target analysis, as well as any additional guidance received from the President and the Secretary of Defense, determine priorities of systems or subsystems to be targeted (Table 2-3).

Table 2-3. Sample strategic CARVER matrix application

Target Systems	C	A	R	V	E	R	Total
Bulk Electric Power	5	3	3	5	5	5	26*
Bulk Petroleum	5	3	5	4	3	5	25*
Water Supply	3	5	3	5	5	3	24*
Communication Systems	3	4	5	2	2	2	18
Air Transportation	1	1	3	1	2	2	10
Ports and Waterways	1	1	3	1	1	1	8
Rail Transportation	2	4	4	1	4	3	18
Road Networks	1	5	3	5	2	5	21
* Indicates target systems suitable for attack. In this example, the Bulk Electric Power target system has been selected.							

OPERATIONAL CARVER EVALUATION CRITERIA

2-40. The purpose of operational target analysis is to determine the critical subsystem or component within the strategically critical system identified for interdiction (Table 2-4). This analysis is done by units at the battalion level and below. Table 2-4 is a notional product resulting from analysis that could be conducted by an SFODA or an SF battalion staff.

Table 2-4. Sample operational CARVER matrix application

Target Subsystems	C	A	R	V	E	R	Total
Generation	5	3	4	3	5	4	24*
Transmission	2	5	2	5	2	5	21*
Control	3	1	4	1	3	3	15
Distribution	2	4	2	4	2	3	17
* Indicates target subsystems suitable for attack. In this example, the Bulk Electric/Generation subsystem has been selected.							

TACTICAL CARVER EVALUATION CRITERIA

2-41. The purpose of tactical target analysis is to determine the military importance of target components, the priority of attack, and the weapons required to obtain a desired effect on a target or set of targets within a target system. Tactical-level analysis lists the complexes or the components of subsystems or complexes selected for attack (Table 2-5).

Table 2-5. Sample tactical CARVER matrix application

Target Component	C	A	R	V	E	R	Total
Water Intake	3	5	1	1	5	4	19
Water Filters and Pumps	5	4	5	4	5	3	26*
Ion Filter	2	1	1	1	5	1	11
Preheater and Pumps	5	2	4	3	5	2	21*
Air Intake	2	1	1	1	5	1	11
Blowers	2	2	1	1	5	1	12
Barges	1	5	1	4	1	5	17
Docks and Oil Pumps	3	5	2	3	1	4	18
Storage Tanks	1	4	1	4	1	5	16
Preheaters and Pumps (Fuel)	5	4	4	3	5	4	25*
Boiler	5	4	5	3	5	4	26*
Turbine/Generator	5	3	5	4	5	5	27*
Transformers	3	4	2	4	5	4	22*
Power Lines	5	1	1	1	1	1	10
Switching Station	2	1	1	2	1	1	8
* Indicates target components suitable for attack. In this example, the Bulk Electric/Generation/Turbine target has been selected.							

2-42. As each potential target is evaluated for each CARVER factor, the analyst enters the numerical rating into the matrix. When all the potential targets have been evaluated, the analyst adds the scores for each target. The totals represent the relative desirability of each potential target and constitute a prioritized list of targets. The targets with the highest totals are considered first for attack.

2-43. If additional personnel or resources are available, the analyst allocates those to the remaining potential targets, in descending numerical order. This allocation scheme maximizes the usable resources. The CARVER matrix gives operational planners a variety of attack options. With the matrix, the analyst can discuss the strengths and weaknesses of each COA against the target (the essence of a feasibility assessment). Drawing conclusions through a sound evaluation process, the analyst can defend his choices before his commander and staff. The analyst should—

- Consider the evaluation criteria carefully.
- Select measurable, quantifiable criteria to the maximum extent possible.
- Apply the criteria consistently.
- Be objective. (The CARVER evaluation system becomes irrelevant if subjectivity is allowed in the process.)
- Consider including a brief explanation of how values or conclusions were reached.
- Consider the top three or four target systems or components during planning. Often, the matrix totals are close or even tied. Selecting the top systems allows for the selection of primary, alternate, and contingency planning.

TIME-SENSITIVE TARGETING

2-44. A target is time-sensitive when it requires an immediate response because it is either a highly lucrative fleeting target of opportunity or it is, or soon will be, a danger to friendly forces. A mission is time-sensitive when there is a limited window of opportunity during which the objective of the mission must be or can be attained. A target may transition from one of low or moderate value to one that is of high value only at a specific time.

2-45. Army special operations units conduct clandestine reconnaissance, surveillance, terminal guidance, and control of weapons systems against time-sensitive targets. Their intelligence staffs may be called upon to provide intelligence support for direct action missions against time-sensitive targets. In such instances, three general sets of planning circumstances frame the initial intelligence available to the staff:

- A known target previously available for attack or reconnaissance by special operations assets over a long period.
- A target on which limited to moderate intelligence already is available within an established operational area in which special operations units are designated as a quick-reaction force or specifically designated to interdict time-sensitive targets.
- A previously unknown time-sensitive target.

Units tasked to destroy or disable a time-sensitive target require detailed intelligence to plan an operation that will not compromise their primary mission and require extraction of the team.

2-46. Time sensitivity can play an important part in categorizing a target and determining its appropriateness as a special operations target. Time sensitivity can be viewed from either a targeting or mission planning perspective or a combination of both, as in the case of personnel recovery missions. Time-sensitive targets invariably require some compression or truncation of the phases of the targeting cycle. Typically, Phases I through IV of the targeting cycle are compressed into the commander's time-sensitive target guidance. Commanders may establish standing operating procedures for facets of time-sensitive target mission planning; for instance, specific go/no-go criteria for the use of ARSOF in time-sensitive target interdiction may be established.

2-47. Compression of the targeting cycle involves a rapid but thorough risk assessment. Particular time-sensitive targets may pose such a high threat to personnel or to mission accomplishment that commanders are willing to accept a higher level of risk in order to attack the target immediately upon detection. Normally, the risk associated with time-sensitive targets involves the possible trade-off of diverting SOF assets from another mission. Commanders and planners must weigh the risks involved and balance the time required for proper planning and execution against the danger of not engaging the target in time. Risk tolerance guidelines may exist as stand-alone documents, such as time-sensitive target standing operating procedures. Preapproved disseminated risk guidelines and approval authorities allow planners to continue planning without continual commander's guidance. This method provides maximum efficiency in a time-sensitive target scenario.

2-48. Once the decision to engage a time-sensitive target has been made, intelligence staffs support planning during Phase V of the time-sensitive targeting process. Modified from the normal joint targeting cycle (Figure 2-3, page 2-14) to aid timely target interdiction, Phase V is composed of the following subphases for time-sensitive targets:

- Find.
- Fix.
- Track.
- Target.
- Engage.

2-49. The find, fix, and track subphases of Phase V are reconnaissance and surveillance asset intensive. Intelligence support to the compressed time-sensitive targeting process requires early identification of priority intelligence requirements to provide focused collection, analysis, and production. Critical information for special operations units interdicting time-sensitive targets includes, but is not limited to—

- Infiltration/exfiltration route data.
- Structural/technical data on target systems, subsystems, and components.
- Enemy force composition.
- Weather/light data.

Chapter 2

- GEOINT.
- Movement routes/patterns of mobile time-sensitive targets.

2-50. Combat assessment is identical in both time-sensitive and regular targeting. Assessment of the engagement of time-sensitive targets is reconnaissance and surveillance asset demanding as well. Intelligence support to time-sensitive target engagement assessment includes not only the immediate (first-order) effects achieved, but the second- and third-order effects resulting as well. Generally, Army special operations units are not used to engage time-sensitive targets with less than operational-level impact and their employment is reserved for engaging time sensitive targets with strategic value. National or theater-level intelligence assets are often tasked to assess the effects of special operations engagement of these high-value targets. Use of national or theater-level assets does not preclude the use of the special operations unit's reconnaissance and surveillance assets.

	Correlation Between the Joint Targeting Cycle Phases and the Time-Sensitive Targeting Process Phases	
	Joint Targeting Cycle Phases	**Time-Sensitive Targeting Process Phases**
Correlation	I. Commander's Objectives, Guidance, and Intent	Commander's Time-Sensitive Target Guidance
	II. Target Development, Validation, Nomination, and Prioritization	
	III. Capability Analysis	
	IV. Commander's Decision and Force Assignment	
	V. Mission Planning and Force Execution • Detect • Locate • Identify • Decide • Strike	• Find • Fix • Track • Target • Engage
	VI. Combat Assessment • Assess	• Assess

Figure 2-3. Time-sensitive targeting

FIND, FIX, FINISH, EXPLOIT, AND ANALYZE TARGETING METHODOLOGY

2-51. The find, fix, finish, exploit, and analyze (F3EA) cycle is a methodology for SOTFs executing precision strike operations at the operational level (Figure 2-4, page 2-15). Critical to the success of the exploit phase are sensitive site exploitation with follow-on document exploitation and target exploitation.

Intelligence Support to Targeting

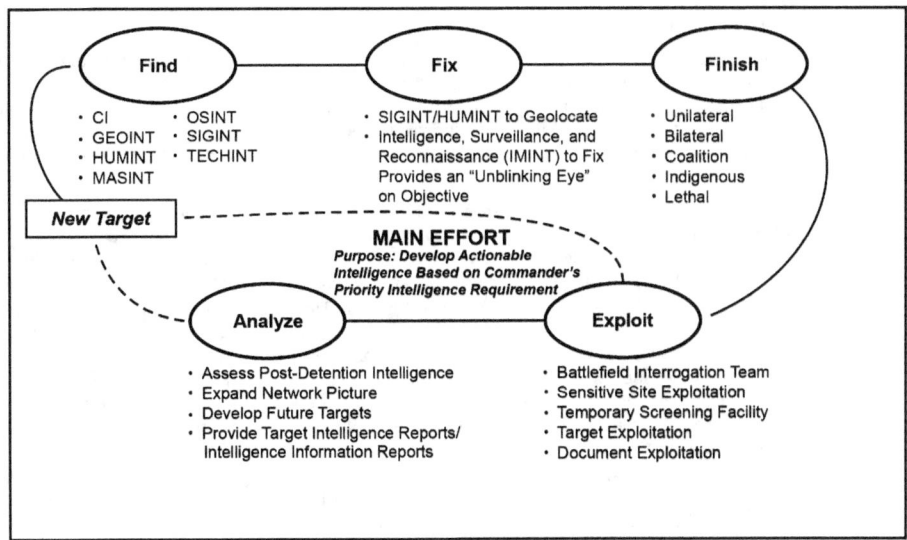

Figure 2-4. F3EA targeting methodology

COMBAT ASSESSMENT

2-52. Combat assessment is the determination of the overall effectiveness of force employment during military operations. Combat assessment is done at the tactical level of war. It is composed of three major components: battle damage assessment, munitions effectiveness assessment, and reattack recommendation. Intelligence production support for combat assessment includes detailed assessments of damage to the adversary's facilities and combat capability, summaries of adversary actions, predictions of adversary intent, analysis of collateral damage, and recommendations for future operations. Combat assessment is the final phase (VI) of the joint targeting process and is defined as the determination of the overall effectiveness of force employment during military operations. Combat assessment helps commanders understand how current operations are progressing and how future operations might best be shaped. Combat assessment answers three fundamental questions:

- Battle damage assessment—Were the desired outcomes (effects) achieved at the target and with respect to the larger target system?
- Munitions effectiveness assessment—Did the forces assigned perform as expected?
- Reattack recommendation—If the desired outcomes were not achieved, or if the employed forces did not perform as expected, what should be done now?

2-53. At the joint level, the operations staff is responsible for supervising combat assessment. Conducting combat assessment requires the analysis and reconnaissance and surveillance capabilities of intelligence staffs. In land component targeting, assessment is typically conducted with identical staff relationships and responsibilities. Intelligence Soldiers are critical to leveraging multiple intelligence disciplines to—

- Determine the effects achieved by current operations.
- Support future operations through supporting development of lessons learned.
- Continue the intelligence process on any targets determined to require reengagement.

Chapter 2

BATTLE DAMAGE ASSESSMENT

2-54. Battle damage assessment is a timely and accurate estimate of damage or degradation resulting from the application of military force, either lethal or nonlethal, against a target. For battle damage assessment to be useful, mission objectives must be observable, measurable, and obtainable. During combat, battle damage assessment reporting should follow standardized formats and timelines, and be passed to command planners and force executors immediately. JP 3-60, *Joint Targeting*, details TTP for planning, conducting, and reporting battle damage assessment.

2-55. Joint command intelligence staffs are responsible for conducting battle damage assessment. They establish battle damage assessment cells with appropriate representation. Subordinate or adjacent Army special operations units may provide representatives to provide subject-matter-expert support on both lethal and nonlethal effects. Battle damage assessment consists of the three phases described below:

- *Phase 1—Physical Damage Assessment.* A physical damage assessment is an estimate of the quantitative extent of physical damage (through munitions blast, fragmentation and/or fire damage effects) to a target element based on observed or interpreted damage. Some representative sources for data that analysts examine to make a physical damage assessment include mission reports, aircraft cockpit video, weapon system video, and visual/verbal reports from ground spotters or troops, controllers, and observers. SIGINT, HUMINT, IMINT, MASINT, and OSINT are used as well.
- *Phase 2—Functional Damage Assessment.* The functional damage assessment is an estimate of the effect of military force to degrade or destroy the functional or operational capability of a target to perform its intended mission. Functional assessments are inferred from the assessed physical damage and all-source intelligence information. This assessment must include an estimation of the time required for recuperation or replacement of the target's function. Battle damage assessment analysts need to compare the original objective for the attack with the current status of the target to determine if the objective was met.
- *Phase 3—Target System Assessment.* Target system assessment is a broad assessment of the overall impact and effectiveness of military force applied against an adversary target system relative to the operational objectives established.

MUNITIONS EFFECTIVENESS ASSESSMENT

2-56. Munitions effectiveness assessment is an assessment of the military force applied in terms of the weapon system and munitions effectiveness to determine and recommend any required changes to the methodology, tactics, weapon systems, munitions, fusing, and/or delivery parameters to increase force effectiveness. Munitions effectiveness assessment is conducted concurrently and interactively with battle damage assessments. Munitions effectiveness assessment is primarily the responsibility of component operations, with inputs and coordination from intelligence. A munitions effectiveness assessment report details weapon performance against specified target types.

REATTACK RECOMMENDATION

2-57. Battle damage assessment and munitions effectiveness assessment provide systematic advice on reattacking targets, which culminates in a reattack recommendation and future targeting, and thus guides further target selection (or target development). Recommendations range from attacking different targets to changing munitions and/or delivery tactics. The reattack recommendation and future targeting must be assessed against the relative importance of the target to the targeting effort or campaign being run.

INFORMATION-RELATED CONSIDERATIONS FOR COMBAT ASSESSMENT

2-58. Information-related capabilities employment methods differ from traditional force application; therefore, targeting analysts performing combat assessment sometimes use different mechanisms to measure the weapons effect on a target and the resultant effect in achieving the objective. Information operations staff sections use a modified form of lethal effects battle damage assessment; however, munitions effectiveness assessment and reattack recommendation that information operations planners use are similar to traditional combat assessment processes. Special operations personnel (primarily CA and MISO) provide input on the modified battle damage assessment by analyzing civil and psychological effects precipitated by lethal effects. The challenge of using the modified battle damage assessment is that second- and third-order effects, and possibly even direct effects from nonlethal fires, may not be readily observable. Therefore, a more detailed analysis of an entire target system or systems may be necessary.

2-59. The methodology for the modified battle damage assessment uses a change assessment, functional damage assessment, and target system assessment to determine the effectiveness of the weapons and tactics employed to achieve the stated objective. Change assessment is based upon observed or interpreted battle damage indicators at selected monitoring points. It uses a systematic understanding of complex target systems, and intelligence capabilities identify and assess changes associated with the target. The quantitative extent of change assessment is used to assess the resulting functional damage.

MEASURES OF EFFECTIVENESS

2-60. Measures of effectiveness are a criterion used to assess changes in system behavior, capability, or operational environment that is tied to measuring the attainment of an end state, achievement of an objective, or creation of an effect. They do not measure task accomplishment or performance. These measures typically are more subjective than measures of performance, but can be crafted as either qualitative or quantitative. Measures of effectiveness can be based on quantitative measures to reflect a trend and show progress toward a measurable threshold.

2-61. All objectives should have one or more associated measures of effectiveness. Although collected upon during operations, they are developed during the first phases of the targeting cycle. To be useful as a gauge of combat effectiveness, a measure of effectiveness must be relevant, measurable, responsive, and resourced. A relevant measure of effectiveness must accurately express the intended effect. An observable measure of effectiveness must be measurable by existing intelligence collection methods. Intelligence supports measure of effectiveness development by quantifying a target's CARVER factors and by identifying observable measures of effectiveness.

MEASURES OF PERFORMANCE

2-62. Measures of performance are criteria for measuring task performance or accomplishment. Measures of performance are generally quantitative, but also can apply qualitative attributes to task accomplishment. They are used in most aspects of combat assessment—since it typically seeks specific quantitative data or a direct observation of an event to determine accomplishment of tactical tasks—but have relevance for noncombat operations as well (for example, tons of relief supplies delivered or noncombatants evacuated). Measures of performance can also be used to measure operational and strategic tasks, but the type of measurement may not be as precise or as easy to observe.

This page intentionally left blank.

Chapter 3
Army Special Operations Forces Intelligence Support Systems and Architecture

This chapter focuses on Army special operations unit organic intelligence systems and architecture and select joint, national, and Department of the Army (DA) systems and architecture. Information on multinational intelligence support is in Appendix C. The interface between Army special operations unique systems and the migration of data gathered from Army special operations unique systems to other systems is a perennial challenge to intelligence staffs. The current methods, systems, and organizations for meeting this challenge are covered in this chapter.

CONNECTIVITY, SYSTEMS, AND ARCHITECTURE

3-1. The global information grid can be described as the sum total of capabilities, processes, and personnel that collect, process, store, disseminate, and manage information for commanders, policy makers, and support personnel in a globally interconnected environment. The global information grid includes all communications and computing systems, software, data, security services, and other associated services owned or leased by the DOD. All Army special operations systems and architecture are subsequently part of the global information grid.

3-2. The intelligence portion of the global information grid is the DOD Intelligence Information System. The DOD Intelligence Information System is the aggregation of personnel, procedures, equipment, computer programs, and supporting communications of the intelligence community. The DOD Intelligence Information System defines the standards for intelligence system, application interoperability, and system compatibility. This program includes hard-copy products, digital or "soft-copy" products, Internet and intranet access to databases, and the ability to "push" or "pull" files of information between producers and consumers. This is accomplished by using data storage, document imaging, electronic publishing, and internal or external networks.

3-3. Army special operations intelligence processes rely on a communications backbone consisting of Joint Worldwide Intelligence Communications System (JWICS) and SECRET Internet Protocol Router Network (SIPRNET). Within this architecture are several systems that support individual or multiple intelligence disciplines. A complete list of intelligence systems in the global information grid is as follows:

- JWICS.
- Joint Deployable Intelligence Support System (JDISS).
- Global Broadcast Service/Integrated Broadcast System.
- Defense Message System.
- Global Command and Control System-Integrated Imagery and Intelligence.
- Joint Intelligence Virtual Architecture.
- Intelligence link (INTELINK).
- Requirements management system.
- Collection Management for Mission Applications.
- Community On-Line Intelligence System for End-Users and Managers (COLISEUM).
- Web Secure Analyst File Environment.

Chapter 3

- Modernized integrated database (MIDB).
- PORTICO.

3-4. Joint interoperability, streamlined flow of information, and pull-down of intelligence tailored to the needs of Army special operations units are key to successful intelligence systems support. For special operations units, increasing connectivity is achieved through the use of organic communications systems. Bandwidth requirements to send and receive many intelligence products can be prohibitive without an organic means of transmission. Systems for dedicated transmission of intelligence data are detailed below. Whenever possible, Army special operations unit intelligence staffs deploy with dedicated intelligence data transmission capability.

ORGANIC INTELLIGENCE SYSTEMS

3-5. The following paragraphs detail the systems currently used by Army special operations units to both collect intelligence and move it through the intelligence process. Systems used by Army special operations personnel include systems common to the Army and unique to Army special operations units.

JOINT DEPLOYABLE INTELLIGENCE SUPPORT SYSTEM-SPECIAL OPERATIONS COMMAND, RESEARCH, ANALYSIS, AND THREAT EVALUATION SYSTEM

3-6. The Joint Deployable Intelligence Support System-Special Operations Command, Research, Analysis, and Threat Evaluation System (JDISS-SOCRATES) program is a garrison sensitive compartmented information (SCI) intelligence automation architecture directly supporting the command's global mission by providing a seamless and interoperable interface with special operations, DOD, national, and Service intelligence information systems. It provides the capabilities to execute the mission command warfighting function, planning, collection, collaboration, data processing, video mapping, a wide range of automated intelligence analysis, intelligence dissemination, imagery tools and applications, imagery dissemination, and news and message traffic applications. The program ensures intelligence support to mission planning and the IPB by connecting numerous data repositories while maintaining information assurance. The JDISS-SOCRATES local area network/wide area network provides single-terminal access to JDISS hosts. The local area network/wide area network is extended via the special operations telecommunications system (SCAMPI) and JWICS. This provides connectivity to information sources at home station and while deployed, as well as connecting garrison-based assets to deployed units, facilitating support to deployed units. Units with imagery analysts are fielded the Enhanced Imagery Workstation (EIW) under the Special Operations Command, Research, Analysis, and Threat Evaluation System (SOCRATES) program. The EIW provides a unit a significant imagery exploitation and GEOINT system production capability. The system is fielded down to the battalion level.

MAN-TRANSPORTABLE SOCRATES-METEOROLOGICAL AND OCEANOGRAPHIC SYSTEM

3-7. The Man-Transportable SOCRATES-Meteorological and Oceanographic System (MTS-METOC) is a briefcase-sized highly mobile, modular analysis system that gives deployed personnel at the SOTF and battalion level the capability to download satellite weather data through organic communications to provide weather support. Each workstation, printer, and external drive is mounted in a ruggedized transit case and weighs less than 30 pounds.

TACTICAL LOCAL AREA NETWORK

3-8. The tactical local area network (TACLAN) program includes segments for the primary communications networks (Non-Secure Internet Protocol Router Network [NIPRNET], SIPRNET, and JWICS). The TACLAN provides Army special operations units with automated intelligence product generation and dissemination capabilities. The TACLAN provides the tactical extension of the garrison JDISS-SOCRATES environment. The TACLAN has been fielded to USAJFKSWCS, SF, Rangers, SOA units, CA brigades, and MIS battalions. The TACLAN is used to receive, process, and manipulate unfinished near-real-time intelligence information as well as information resident in national and theater intelligence databases. The TACLAN software baseline should be the same as the JDISS software baseline for analyst operations and be configured to accept theater-unique software applications. The TACLAN can interface with fixed or mobile communications equipment to send and receive data, and record communications. This capability allows intelligence analysts to search national, theater, and Service intelligence sites for intelligence data in support of mission requirements. It also allows users to receive unfinished near-real-time information and process/manipulate it into usable intelligence in a field environment. It is designed to rapidly disseminate critical intelligence between Army special operations elements, conventional forces, intelligence facilities, and national intelligence agencies.

ENHANCED IMAGERY WORKSTATION

3-9. The EIW is a specialized JDISS-SOCRATES workstation providing imagery analysts workstations with tailored applications and peripheral hardware to meet USSOCOM imagery receipt, exploitation, production, and dissemination requirements. There are desktop and laptop configurations allowing imagery analysts to meet garrison and tactical needs to receive and send unexploited and exploited GEOINT and imagery products directly to users. EIWs are fielded to each SF group, the Ranger Regiment, and the SOAR. The EIW provides the following capabilities for these select units to meet GEOINT requirements:

- Access to imagery and equipment able to receive and disseminate GEOINT products.
- Imagery manipulation and the capability to generate hard- and soft-copy products.
- The capability to store and archive GEOINT products.

Included with the EIW are a large format plotter and a color printer. EIW software includes Earth Resources Data Analysis System (ERDAS) imagery manipulation modules, image exploitation support system interface, electronic light table 5500 and Remote View, imagery product library, and the requirements management system.

BRIEFCASE MULTIMISSION ADVANCED TACTICAL TERMINAL AND MULTIMISSION ADVANCED TACTICAL TERMINAL

3-10. The Briefcase Multimission Advanced Tactical Terminal (BMATT) is a stand-alone, briefcase-sized, highly mobile Integrated Broadcast System tactical data receiver. The Multimission Advanced Tactical Terminal (MATT) is configured for vehicle mounting. The BMATT provides near-real-time interface with global and theater intelligence threat warning broadcasts without greatly increasing space and weight requirements. The system is fielded to select units. The USSOCOM embedded national tactical receiver will replace the MATT and BMATT.

OMNISENSE II-E

3-11. The OmniSense II-E is a remote intrusion detection sensor system developed by the United States Army Intelligence and Security Command (INSCOM) for global operations against terrorism networks to replace the improved-remotely monitored battlefield sensor system as a bridge to the future force unattended sensor. It provides a persistent route, area, and point surveillance capability. It detects targets autonomously, using acoustic, magnetic, passive infrared, and seismic sensors. It images targets with infrared or visual cameras, and sends information via remote satellite link to theater-wide servers. Sensor data is distributed to remote users via secure Web sites and the Integrated Broadcast System.

DISTRIBUTED COMMON GROUND SYSTEM-ARMY

3-12. The Distributed Common Ground System-Army (DCGS-A) is the reconnaissance and surveillance component of the Army's Battle Command System and the primary system for reconnaissance and surveillance tasking, posting, processing, and using information about the threat, weather, and terrain at the JTF level and below. The DCGS-A provides the software applications necessary for commanders to access information beyond what is collected by organic assets. The DCGS-A is the reconnaissance and surveillance gateway to joint, interagency, multinational, and national data; information; intelligence; and collaboration. It provides access to theater and national intelligence collection, analysis, early warning, and targeting capabilities in support of Army units. The DCGS-A synchronizes reconnaissance and surveillance tasking, posting, processing, and using efforts; operates in a networked environment at multiple security levels; and supports the intelligence reach and split-based operations to improve accessibility to data and reduce the forward footprint. DCGS-A software and hardware provide a single integrated reconnaissance and surveillance ground processing system composed of joint common components that are interoperable with sensors, other information sources, all warfighting functions, and the DOD DCGS Family of Systems. DCGS-A software and hardware is tailored by echelon and scalable to the requirements of each mission, task, and purpose.

ALL-SOURCE ANALYSIS SYSTEM-LIGHT, AN/TYQ-93(V)4

3-13. The All-Source Analysis System-Light (ASAS-L) provides a lightweight, tactical intelligence system capable of interfacing with Army Battle Command System networked systems using a variety of radio and network protocols. The ASAS-L incorporates many of the same features as the DCGS-A in a notebook computer. The ASAS-L can accomplish the direct exchange of map graphics and overlays with DCGS-A. The ASAS-L uses the Joint Mapping Tool Kit as a mapping package, supports standard All-Source Analysis System (ASAS) all-source correlated database and custom, user-defined, database capability. The ASAS-L provides the intelligence analyst with some IPB capabilities, such as manual templating, modified combined obstacle overlay development, and overlay capabilities for COA development. The ASAS-L provides viewing and limited imagery annotation capability with both commercial off-the-shelf and Government off-the-shelf software tools and a standard suite of office automation applications.

ALL-SOURCE ANALYSIS SYSTEM-SINGLE SOURCE, AN/TYQ-52(V)

3-14. The All-Source Analysis System-Single Source (ASAS-SS) is a component of ASAS Block I and provides SIGINT analysts tools capable of sophisticated analytic processing and reporting. The ASAS-SS receives SCI SIGINT and processes it into multidiscipline intelligence products. In support of this, the ASAS-SS is capable of automatically receiving and processing messages from national to tactical collectors and passing relevant data into local databases. The ASAS-SS provides a comprehensive set of ELINT, communications intelligence, and intelligence correlation analysis tools, as well as national database browsers. It supports the generation of messages in multiple standard formats. The system is fielded to select SF groups.

JOINT TACTICAL TERMINAL-BRIEFCASE, AN/USQ-161(V)1(C)

3-15. The Joint Tactical Terminal-Briefcase (JTT-B) allows Army special operations units to receive the Integrated Broadcast System, a series of secure intelligence broadcasts, to support operations against likely targets, as well as to quickly develop and tailor new databases to unexpected threats. The systems deliver critical time-sensitive battlefield targeting information to tactical commanders and intelligence nodes at all echelons, in near-real time at the collateral level. The JTT-B will support the intelligence analyst in the areas of indications and warnings, situation assessment, target analysis, mission planning and rehearsal, imagery analysis, and electronic support measures. Additionally, it will provide extensive capability to manipulate data received via the Integrated Broadcast System and to overlay these products with database information. It also provides word processing, message text generation, and local network services. The JTT-B can receive and display reports from units and other joint blue force tracking and situation awareness devices. The USSOCOM embedded national tactical receiver will replace the JTT-B.

COUNTERINTELLIGENCE/HUMAN INTELLIGENCE AUTOMATED TOOL SET, AN/PYQ-3(V)3

3-16. The Counterintelligence/Human Intelligence Information Management System (CHIMS) or the Counterintelligence/Human Intelligence Automated Tool Set (CHATS), or a variant, is a team leader device that interfaces with the DCGS-A, the Counterintelligence and Interrogation Operations workstation, and the Individual Tactical Reporting Tool agents/collector device. The CHIMS/CHATS provides automation capability to collect, manage, receive, store, and export text, electronic data, and digital imagery information, and to prepare, process, and disseminate standard messages. The system is fielded at the SF battalion, SOAR, and Ranger Regiment levels.

Joint Threat Warning System

3-17. The Joint Threat Warning System is a USSOCOM program that migrates legacy SIGINT capabilities used by USSOCOM components into an evolutionary acquisition strategy. The Joint Threat Warning System provides threat warning, protection, and enhanced situational awareness information to units through SIGINT exploitation. It provides a plug-and-play software architecture across maritime, air, and ground applications, which allows a customized, scalable approach for each mission from simple radar-warning receiver-type functions in the Body-Worn variant to a fully capable suite of SIGINT systems in the Team Transportable System. The Joint Threat Warning System program includes the Ground Signals Intelligence Kit AN/PRD-14(V)1 consisting of the Body-Worn System and man-portable configurations for specific missions, and Team Transportable, maritime, and air variants.

Individual Tactical Reporting Tool, AN/PYQ-8

3-18. The Individual Tactical Reporting Tool is a hand-held device designed for the individual CI agent or HUMINT collector. The Individual Tactical Reporting Tool provides a notepad and recording capability; report masks; limited mapping capability; a local database; the ability to receive, process, store, and disseminate digital imagery; message masks; and a communications interface with the CHIMS/CHATS. The system is fielded to SF groups, the SOAR, and Ranger units.

Counterintelligence and Interrogation Operations Workstation, AN/PYQ-7

3-19. The Counterintelligence and Interrogation Operations Workstation provides the intelligence staff at all levels with CI and HUMINT automation support for CI and interrogation planning and management operations. Automated support includes CI/interrogation planning tools; communications interfaces for national databases; reporting tools; analysis tools; and asset, source, and mission management tools. The system is fielded to SF groups.

Digital Topographic Support System-Deployable, AN/TYQ-71

3-20. The Digital Topographic Support System-Deployable (DTSS-D) accepts topographic, geospatial, and remotely sensed imagery data from the NGA, Service, commercial source, and user-generated overlays. DTSS-D functional capabilities include the creation of mobility, environmental, three-dimensional terrain visualization, fly-through, and custom geospatial products, as well as the creation, augmentation, modification, and management of topographic data. The DTSS-D provides updated map background and terrain intelligence information to the Army Battle Command System workstations, and accepts terrain intelligence and data updates from these systems, as well as from any user of geospatial data. The DTSS-D also provides data for the common operational picture. The DTSS-D uses commercial and government off-the-shelf technology in printers, scanners, and computer workstations, combined with image processing and geographic information systems software (ArcGIS, ERDAS, and other specialized software). The DTSS-D provides the capability to generate, print, and disseminate geospatial data (maps, custom products, and imagery maps) through a variety of media.

Integrated Meteorological System-Light, AN/GMQ-36

3-21. The Integrated Meteorological System-Light (IMETS-L) is the Army's tactical weather communication, intelligence, and information system providing digital weather support to enable SOWTs to support tactical units. The IMETS-L is interoperable with other Army Battle Command Systems. During mission execution, the IMETS-L helps SOWTs to advise supported units to evaluate what they can see and shoot, and facilitates better communication. To produce digital weather information, the IMETS-L combines advanced observations from weather satellites, upper air information, surface weather observing networks, and numerical models to build a gridded meteorological database in the area of operations. The IMETS-L enhances the tactical commander's situation awareness by providing visualization of weather. This system is fielded to SF battalions, the SOAR, and the Ranger Regiment.

ALL-SOURCE ANALYSIS SYSTEM BLOCK II ANALYSIS AND CONTROL ELEMENT-LIGHT, AN/TYQ-90

3-22. The ASAS Block II Analysis and Control Element-Light uses an intelligence-shared server and All Source Workstations in a server-client relationship to provide a robust processing capability over previous products. The use of the intelligence-shared server allows all of the single-source/single-intelligence disciplines to access the DCGS-A all-source correlated database and various single-source databases stored on the intelligence-shared server via a Web-based interface. This Web interface will provide the capability of performing interactive cross-linking of data from multiple data sources and aid in providing the commander a more accurate depiction of ground truth. In addition to the components listed above, the Block II Analysis and Control Element-Light includes the Multiple-Intelligence Processor, Open-Source Collection Workstation, CI/HUMINT Workstation, Trusted Workstation, and TES Remote Intelligence Server.

TROJAN SPECIAL PURPOSE INTEGRATED REMOTE INTELLIGENCE TERMINAL LIGHTWEIGHT INTEGRATED TELECOMMUNICATIONS EQUIPMENT, AN/TSQ-226(V)1

3-23. The Army Trojan Special Purpose Integrated Remote Intelligence Terminal (SPIRIT) Lightweight Integrated Telecommunications Equipment (LITE), known as the TS-LITE system, provides tactical commanders with high-capacity, near-real-time access to intelligence from Trojan Classic sites, national agencies, and other tactically deployed elements using tactical satellite communications terminals. The TS LITE supports quick-reaction reporting and analysis, and links tactical-to-satellite communications capability for contingency/exercise deployments until the Warfighter Information Network-Tactical is completely fielded. The system is fielded to SF groups, the SOAR, and Ranger units.

BROADCAST REQUEST IMAGERY TECHNOLOGY ENVIRONMENT

3-24. The NGA-sponsored Broadcast Request Imagery Technology Environment (BRITE) program is a computer-based GEOINT information system enabling deployed users the ability to exploit high-resolution national satellite and theater airborne reconnaissance imagery, at the Secret level, within minutes of collection. BRITE can be employed using satellite broadcast or network communications links. BRITE View software and supporting intelligence architecture enables operators to receive and display notifications (footprints) of the current imagery collected in the area, then select an area (chip) within the footprint of an image of interest and command the system to reach back to the server and pull the chip of imagery from the image of interest. The server automatically pushes the image chip to the user using the user's specified dissemination path. The chip is delivered to the user in a standard NGA national imagery transmission format that can be exploited with the resident imagery exploitation applications on JDISS-SOCRATES and TACLAN. The 75th Ranger Regiment and the 160th SOAR have Global Broadcast Service-based BRITE broadband receive suites with low probability of intercept and low probability of detection communications devices. The BRITE View software is included on TACLAN and JDISS-SOCRATES, enabling BRITE SIPRNET access.

LOW-COST S-BAND RECEIVER

3-25. The Low-Cost S-Band Receiver is a TACLAN component that permits receipt of the Automated M-22 Broadcast. The Automated M-22 Broadcast supports disadvantaged users and provides an alternate means to receive intelligence products. This network operates at up to 256 kilobits per second. The enhanced tactical broadcast adds a low probability of intercept and low probability of detection reachback communications device to the Low-Cost S-Band Receiver that enables intelligence-smart push/pull operations. The enhanced tactical broadcast with the BRITE View software provides tactical users a low-bandwidth BRITE capability. The system is fielded to SF groups, the SOAR, and Ranger units.

TACTICAL EXPLOITATION SYSTEM-LOW-COST INTELLIGENCE TACTICAL EQUIPMENT, Z00677

3-26. The Tactical Exploitation System-Low-Cost Intelligence Tactical Equipment (TES-LITE) combines the Tactical Exploitation of National Capabilities program functionality into a single integrated, downsized, and scalable system, capable of split-based or collocated operations. The TES-LITE serves as the interface between national collectors and in-theater operating forces, and also directly receives data from theater and tactical sensors. The TES-LITE receives, processes, exploits, and disseminates data from direct downlinks and ground stations to tactical, national, and theater platforms. The TES-LITE's advanced communications architecture can simultaneously receive multiple broadcasts for operations and product dissemination in a modular, interoperable configuration to support tactical and national imagery requirements. It interfaces with, and serves as the preprocessor for, the ASAS, DCGS-A, Joint STARS Common Ground Station, and the Digital Topographic Support System. The TES-LITE is a bridging technology to the future lightweight DCGS-A. The system is fielded to SF groups, the SOAR, and Ranger units.

ARMY SPECIAL OPERATIONS FORCES DATABASES AND TOOLS

3-27. There are several current intelligence systems and databases accessible via the JDISS-SOCRATES. Some of these include the following:

- *INTELINK*. INTELINK (Figure 3-1, page 3-9) is the principal electronic means for intelligence product dissemination. INTELINK builds on ongoing architectural initiatives at the Top Secret and/or SCI, Secret, and Unclassified classification levels. INTELINK provides a comprehensive set of tools to query, access, and retrieve information. INTELINK permits collaboration among analysts and users, and will simplify access to a variety of services. The unit's intelligence officer assesses the availability of INTELINK access among assigned and en route forces.

- *Special operations debrief and retrieval system (SODARS)*. The SODARS program was developed to automate and store after-action reports and debriefing reports from special operations elements that conducted missions outside the continental United States (OCONUS). This information is used for future planning and in preparing personnel for deployments. The SODARS program is managed by the J-2, USSOCOM.

- *Modernized integrated database*. The MIDB has been designated by the DOD as the migration system for the production and analysis of the general military intelligence database. The MIDB provides a controlled set of common data elements and applications that will permit rapid, accurate exchange and analysis of intelligence information at production centers, joint commands, Service components, and tactical units.

- *Psychological Operations Automated System-Message Center (POAS-MC)*. This system is located within the USASOC SharePoint environment. The POAS-MC provides a centralized storage location for incoming and archived message traffic, with search capabilities for day-to-day intelligence and/or MISO country studies. This database system provides support at the tactical, operational, and strategic levels, with an automated capability to support state-of-the-art planning, implementation, and evaluation of U.S. MISO. The POAS-MC is accessible through the USASOC SIPRNET portal and requires registration to log in.

- *Cultural Intelligence Special Operations Assessments Web Server*. This server resides on the SIPRNET. It provides the intelligence community access to the MISO studies that are produced by the analysts of the cultural intelligence section of the MISOC(A). The home page is https://poasweb.poas.socom.smil.mil/whats_new/index.htm. The MISO studies are then carried over to the JWICS, which is cleared up to the Top Secret and SCI levels.

Chapter 3

- *PROSPECTOR portal.* The PROSPECTOR portal resides on the SIPRNET and requires an INTELINK account to log in. The PROSPECTOR portal is used to collaborate and synchronize USSOCOM MISO globally. The portal provides three specific functions—
    - For coordination and approval of programs and activities by deployed MISO forces, ensuring transparency between support to the GCC, USSOCOM, and the Joint Staff/Office of the Secretary of Defense.
    - To consolidate and analyze measures of effectiveness from the operational to the strategic level and across the geographic combatant command, as necessary.
    - To serve as a focal point for the collection, collaboration, and sharing of documents, information, and ideas.

    The portal provides the interagency a single location to post and update planning products, a combined calendar of relevant events, a contact list of the specific community-of-interest members, links to other related sites, and a discussion forum to assist in plan development.

- *Imagery product library.* This system is an integrated, unified, soft-copy library system that supports the rapid storage, retrieval, and dissemination of digital data, sponsored by the NGA. It allows users, such as intelligence analysts, imagery analysts, cartographers, and operation personnel, seamless access to the imagery, image-based products, and relevant metadata that they need to fulfill their respective missions. The imagery product library provides the server software for implementing this objective and equips the imagery community with improved accessibility, operational support, and distribution of geospatial and imagery products. The imagery product library achieves its mission by providing an automated capability to support the following activities:
    - Receive imagery and image-based products from multiple sources.
    - Maintain a database of imagery and image-based products.
    - Transfer imagery and image-based products to imagery clients from imagery sources and to remote locations using several formats and compression ratios.

- *Asymmetric Software Kit (ASK).* The ASK provides tools for Soldiers to develop and manage information and intelligence. Data mining is performed using OrionMagic and Pathfinder (found on the SIPRNET). OrionMagic has a built-in Web browser and allows the user to organize information collected in an electronic filing system in the form of cabinets, outlines, folders, and note cards. Pathfinder for the Web is an intelligence data mining and visualization application developed to assist analysts in the processing, searching, and analyzing of large quantities of text data. Analysis and visualization are conducted using Analyst's Notebook and ArcGIS. Analyst's Notebook is used for link analysis. The targeting process can be aided and refined through link analysis using Analyst's Notebook. This visualization of analysis easily identifies links, patterns, and connections that should explain relationships and correlations between people, incidents, and associated groups, and other useful information that can be associated. Further visualization can be conducted using ArcGIS. Using various programs in the ArcGIS suite, all requisite information can be represented geospatially. The ArcGIS enables spatial information created to be queried as an aid in planning and targeting. The ASK also includes a standardized and formatted database for ingesting geospatially referenced data from the field to aid in the dissemination of data to any database user.

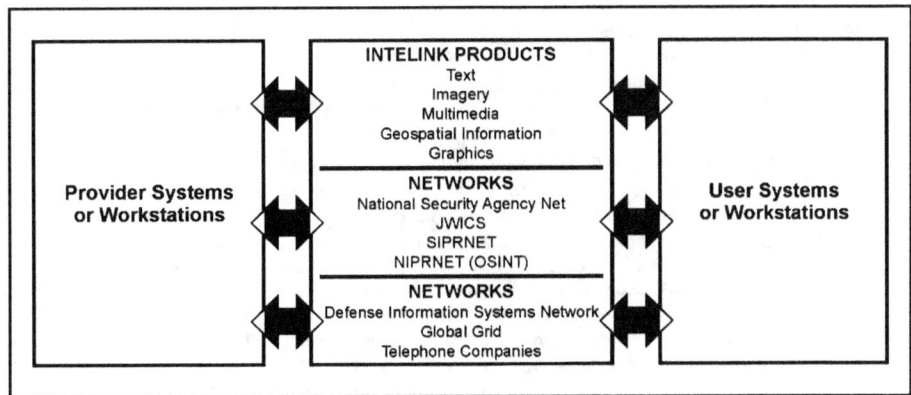

Figure 3-1. INTELINK

ARMY SERVICE COMPONENT COMMAND INTELLIGENCE SUPPORT

3-28. USASOC is the Army Service component command for special operations. USASOC is USSOCOM's Army component and consists of Regular Army and National Guard units. USASOC provides trained and ready SF, Ranger, SOA, Regular Army MISO, and Regular Army CA forces to GCCs and U.S. Ambassadors. USASOC develops unique special operations doctrine, tactics, techniques, and materiel support.

UNITED STATES ARMY SPECIAL OPERATIONS COMMAND INTELLIGENCE STAFF

3-29. The intelligence staff advises and assists the commanding general and his staff on all intelligence and security matters. The intelligence staff provides deliberate and crisis-action intelligence support to all Army special operations units based in the continental United States (CONUS), including collection, requirements management, analysis, production, and dissemination. It provides intelligence and security support to assigned units preparing to conduct missions or exercises. The intelligence staff is divided into two divisions—the Intelligence Operations Division and the Security and Support Division. The major functions of the intelligence staff are as follows:

- Direct intelligence, CI, security, geospatial information and services (GI&S), and weather activities supporting Regular Army special operations units and National Guard SF groups.
- Interface with USSOCOM; Headquarters, Department of the Army (HQDA); USAJFKSWCS; and the U.S. Army Intelligence Center of Excellence to identify, refine, and resolve Army intelligence and Army special operations doctrine, training, equipment, and personnel requirements within USASOC.
- Establish, facilitate, and maintain planning, programming, and policy guidance for long-range intelligence plans, objectives, and architectures.
- Develop and implement policy and exercise oversight for all intelligence automation systems within USASOC and its component subordinate commands and units.
- Develop, implement, and manage the USASOC intelligence oversight program.

Chapter 3

INTELLIGENCE OPERATIONS DIVISION

3-30. The Intelligence Operations Division provides current and estimative intelligence and threat support to Army special operations units. Other major functions include the following:

- Serves as point of contact for all facets of deliberate and crisis-action intelligence support to USASOC HQ and CONUS/OCONUS-based Army special operations units, to include—
  - Collection/requirements management.
  - Dissemination and automated information system support.
  - Select intelligence equipment acquisition, analysis, and production.
  - Personnel augmentation (of the crisis action center or in-theater task forces).
- Conducts continuous and direct liaison with USSOCOM, the National Security Agency (NSA), the Defense Intelligence Agency (DIA), the NGA, the Central Intelligence Agency (CIA), the DOD, the DA, geographic combatant commands, and other government agencies to monitor global intelligence management policy and procedures.
- Coordinates and manages selected Army special operations intelligence special access programs.
- Registers, validates, and processes all CONUS SIGINT, GEOINT, HUMINT, OSINT, MASINT, and CI all-source requirements from component subordinate units, and monitors and deconflicts component subordinate unit OCONUS theater-related information requirements.
- Reviews, validates, and posts to the USSOCOM server Army special operations units' input to the SODARS.
- Reviews and validates Army special operations intelligence concepts of operation.
- Develops policies and implements directives for intelligence support to the Army Service component command.
- Prepares special operations intelligence input to DOD and DA plans, studies, and products; coordinates with USSOCOM, DA, and U.S. Army Training and Doctrine Command proponents.
- Provides predeployment threat and situation briefings for select groups and individuals from staffs of USASOC and component subordinate commands and units, Army special operations units, security assistance teams, and personnel deploying after individual terrorism awareness course instruction.
- Manages the Army Service component command Threat Awareness and Reporting Program (TARP) and provides briefings.
- Advises Army special operations units' intelligence staffs on cryptology issues.
- Serves as the USASOC manager for GI&S and imagery software and system requirements, and provides GI&S and imagery-related support to component subordinate units.
- Manages and coordinates environmental services (meteorological, oceanographic, and atmospheric sciences) and related activities for the command.
- Serves as the Command Language Program Manager for linguists assigned to Army special operations units.
- Serves as the Army Service component command intelligence point of contact for Status of Resources and Training System, Unit Status Report, and Joint Monthly Readiness Review requirements.
- Manages the USASOC Tactical Exploitation of National Capabilities and the USSOCOM National System Support to special operations programs.
- Directs and manages the USASOC portions of the SOCRATES program and USASOC-procured intelligence systems.
- Reviews intelligence-related doctrinal manuals from both the DA and the DOD.
- Provides training assistance to USAJFKSWCS.

SECURITY AND SUPPORT DIVISION

3-31. The Security and Support Division is comprised of a security guard force for the USASOC HQ building, information security, personnel security, special security officer, dedicated sensitive activities personnel, and support office. The following are the major functions of the division:

- Directs the technology transfer/disclosure of information to the foreign nationals and contractors program to protect against unauthorized disclosure of classified and controlled unclassified information to foreign nationals.
- Provides direct foreign disclosure/technology transfer support to the command and all component subordinate units.
- Directs information security, intelligence systems security, personnel security, terminal electromagnetic pulse escape safeguard technique (TEMPEST) security, information systems security monitoring, surveillance countermeasures, industrial security, and foreign disclosure.
- Coordinates USASOC personnel security clearance issues with the Central Clearance Facility, Defense Security Services Center, DA, Department of Justice, and Office of Personnel Management.
- Conducts security inspections at component subordinate units to ensure compliance with USASOC, USSOCOM, and HQDA policies and directives.
- Approves procedures for safeguarding, transporting, and controlling communications security and controlled cryptographic items materiel.
- Develops information security policy and directs the program to safeguard classified information within USASOC HQ and its component subordinate commands and units.
- Establishes and administers the command's SCI security management program and ensures the administration of the program within respective component subordinate commands and units in accordance with (IAW) applicable regulations and guidance.
- Evaluates new, modified, and reconfigured construction plans for sensitive compartmented information facilities (SCIFs) for component subordinate units to ensure TEMPEST and physical security compliance.
- Reviews and approves the deployment of tactical SCIFs in CONUS and the establishment of temporary secure working areas by component subordinate units.

JOINT AND THEATER-LEVEL INTELLIGENCE SUPPORT

3-32. The combatant command intelligence staff provides higher echelons and subordinate commands with a common, coordinated, all-source intelligence picture by applying national intelligence capabilities, employing joint force intelligence resources, and identifying and integrating additional intelligence resources. The intelligence staff normally exercises staff supervision over the JIC, develops and manages an optimal collection plan, and matches resources against requirements. The intelligence efforts of Army special operations units at subordinate joint commands are coordinated by the combatant command intelligence staff and all IPB products produced by these units will be forwarded up to the intelligence staff for integration into the overall joint intelligence preparation of the operational environment effort.

3-33. The combatant command JIC is the focal point for the combatant command's intelligence analysis and production effort. Each combatant command possesses a JIC or an equivalent element. For instance, the United States European Command has a joint analysis center. If the organic intelligence center cannot meet the combatant commander's requirements, the JIC forwards a request for information to the National Military Joint Intelligence Center (NMJIC) or to subordinate command levels using the COLISEUM request for information management system. The JIC may also seek to ensure timely support by submitting requests to national agency liaisons to the command.

3-34. There is no "typical" JIC organizational structure, and each organization varies depending on combatant command requirements. However, all JICs have the following responsibilities:
- Meet the intelligence needs of the combatant command and subordinate joint forces.
- Maintain and coordinate the theater collection plan and employment of theater assigned and supporting sensors.
- Develop and maintain databases.
- Develop or validate battle damage assessment.
- Provide continuous indication and warning intelligence assessments.
- Provide intelligence support and augmentation to subordinate joint forces.

3-35. The organizational structure of the subordinate joint force's intelligence element is determined by the JFC based on the situation and mission. However, most situations will require augmentation of joint force intelligence capabilities through the establishment of a JISE under the supervision of the joint force intelligence staff. The JISE is composed of analytical experts and analysis teams that provide services and products required by the JFC, staff, and components, to include the SOTF. Capabilities of the JISE include order of battle analysis, identification of adversary centers of gravity, analysis of adversary command, analysis of adversary control systems, targeting support, collection management, and maintenance of a 24-hour watch. When a JSOTF is established, a subordinate JISE is established.

3-36. The subordinate joint force intelligence staff is responsible for planning and directing the overall intelligence effort on behalf of the JFC. The intelligence staff develops and recommends priority intelligence requirements based on the commander's guidance, identifies shortfalls in intelligence capabilities and submits requests for additional augmentation, and ensures the intelligence needs of the commander and joint force staff are satisfied in a timely manner. Additionally, at the discretion of the commander, the intelligence staff provides administrative support to augmentation forces and the JISE, including personnel, information, and physical security.

3-37. The JFC normally establishes a CI and HUMINT division (J-2X). This concept is designed to integrate HUMINT and CI by combining the HUMINT operations cell with the joint force's CI coordinating authority into the CI and HUMINT division. A J-2X is the HUMINT and CI focal point for the JFC. As the command's tasking authority for HUMINT and CI collection, the J-2X is responsible for the management, coordination, and deconfliction of HUMINT and CI collection within the operational area. The CI and HUMINT division monitors and supports the activities of the joint exploitation centers, maintains the command source registry, deconflicts source matters, and performs liaison functions with external organizations.

DEPARTMENT OF THE ARMY INTELLIGENCE SUPPORT

3-38. The Army deputy chief of staff, G-2, exercises general staff responsibility for all Army TECHINT activities. The Army deputy chief of staff, G-2, forms policies and procedures for scientific and technical intelligence (S&TI) activities, supervises and carries out the Army S&TI program, coordinates DA staff and direct reporting unit requirements for TECHINT, and is responsible for the Army Foreign Materiel Program.

3-39. INSCOM is responsible for peacetime TECHINT operations. HQ, INSCOM, fulfills its responsibilities through its TECHINT oversight function and manages the Army's Foreign Materiel for Training Program and Foreign Materiel Exploitation Program. It provides the interface with strategic S&TI agencies in support of foreign materiel exploitation and organizes, trains, and equips echelons above corps TECHINT organizations during peacetime. TECHINT exploitation within INSCOM is performed by the following elements:
- *National Ground Intelligence Center (NGIC).* HQ, INSCOM, exercises direct operational control over the NGIC. The NGIC produces and maintains intelligence on foreign scientific developments, ground force weapons systems, and associated technologies. NGIC analysis includes, but is not limited to, military communications electronics systems; types of aircraft used by foreign ground forces; chemical, biological, radiological, nuclear, and high-yield

explosives (CBRNE) systems; and basic research in civilian technologies with possible military applications.
- *203d Military Intelligence Battalion.* The 203d Military Intelligence Battalion is a multicomponent unit headquartered at Aberdeen Proving Ground, Maryland, and is the Army's sole TECHINT battalion. This battalion conducts TECHINT collection and reporting in support of validated S&TI objectives and acts as the HQDA executive agent for foreign materiel used for training purposes.
- *902d Military Intelligence Group.* The 902d Military Intelligence Group provides resident offices at most Army installations and may be contacted through Group HQ at Fort Meade, Maryland. The 902d resident offices provide security services and are the central point of contact for TARP. Resident offices receive reports on any potential network intrusions or data collection attempts by unauthorized persons and agencies.
- *United States Army Topographic Engineering Center (USATEC).* The USATEC facilitates superior knowledge of the battlefield and supports the nation's civil and environmental initiatives through research, development, and the application of expertise in the topographic and related sciences. This support includes three-dimensional environment graphics, hydrologic and environmental data, terrain analysis products, and validation of Army geospatial requirements, systems, programs, and activities. The USATEC works closely with USASOC component subordinate commands and units.

NATIONAL-LEVEL INTELLIGENCE SUPPORT

3-40. Army special operations units' intelligence staffs have access to national-level intelligence through databases, through higher HQ, and sometimes through national agency representatives located at the small-unit level; for example, a Federal Bureau of Investigation (FBI) liaison special agent working with a MISO unit.

3-41. National agency representatives support the combatant commands on a full-time basis through representatives. A JTF or JSOTF JISE may have an equal number of national agencies represented as well. Some of these representatives are located full time at the command's HQ. These representatives serve as the intelligence staff advisors and can be employed as directed by the intelligence staff. Examples include DIA analysts augmenting the all-source production center or the NGA element embedded in the JISE supporting the terrain section.

3-42. A Director of Central Intelligence representative from the CIA is assigned to each of the combatant commands to coordinate CIA and other intelligence community support to the command, and to facilitate access to CIA resources. Director of Central Intelligence representatives can also advise and assist the command regarding the secondary and follow-on dissemination of originator-controlled material and Human Intelligence Control System information IAW Intelligence Community Directive 301, *National Open Source Enterprise*. The CIA contributes significant intelligence used in developing strategy, determining objectives, determining deception objectives, planning and conducting operations, and evaluating the effects of operations.

3-43. The DIA maintains defense intelligence support offices (DISOs) at each of the combatant commands, U.S. Forces Korea, and Supreme Headquarters Allied Powers Europe and North Atlantic Treaty Organization (NATO). Each DISO includes a senior DIA intelligence officer, who serves as chief of the DISO and as the personal representative of the DIA director, an administrative assistant, and a varying number of DIA functional intelligence specialists based on the needs of the supported command.

3-44. The typical DISO includes a HUMINT support element consisting of one or more Defense Human Intelligence Service personnel, an intelligence production liaison officer, and a MASINT liaison officer. The DIA Directorate for MASINT and Technical Collection (DIA/DT) MASINT liaison officers help expedite a broad spectrum of MASINT operational support between DIA/DT and the supported command. Some DISOs also have information technology and Joint Intelligence Task Force Combating Terrorism representatives. The DISO organization enhances and expedites the exchange of information between the DIA and the supported command.

3-45. The NSA provides SIGINT, technical support to information operations planning and coordination activities, and information systems security for the conduct of military operations IAW tasking, priorities, and standard of timeliness assigned by the Secretary of Defense.

3-46. The NGA provides representatives to the combatant commands in the form of either NGA support teams, composed of staff officers and imagery and geospatial analysts, or single representatives to select organizations; for example, the NGA maintains a representative at USASOC HQ, as well as a geospatial analyst at each Regular Army SF group, the Ranger Regiment, and the SOAR, and an imagery analyst at the Ranger Regiment and the SOAR. The NGA representative is the central point of contact for all operational and training support from the NGA, to include GEOINT products and emerging concepts, technologies, and procedures. The NGA is responsible for providing responsive imagery, IMINT, and GI&S support, and fusing that data with intelligence from all sources. For example, the NGA might fuse hydrographic, time-sensitive imagery and MASINT into a single product. The NGA representative may also arrange meteorological and oceanographic support from the joint meteorological and oceanographic officer. The NGA also provides the Remote Replication System to select installations and agencies. The Remote Replication System provides printing on demand for geospatial products and duplication services of NGA standard geospatial data. Fort Bragg, North Carolina, has a Remote Replication System at the Geospatial Readiness Facility that supports all Army special operations units.

3-47. The Defense Logistics Agency distributes classified and unclassified NGA standard digital and hard-copy geospatial products to anyone with a DOD activity address code. These products support the intelligence effort, enabling units to develop and display geospatial data on the common operational picture.

3-48. The National Reconnaissance Office provides representatives to the combatant commands in the form of liaison officers and theater support representatives. These National Reconnaissance Office representatives provide technical assistance relating to the capabilities of National Reconnaissance Office systems. Army special operations units will typically access National Reconnaissance Office representatives and service through a JTF, JSOTF, or through conventional unit intelligence staffs.

3-49. National intelligence support teams (NISTs) are comprised of intelligence and communications experts from the DIA, the CIA, the NGA, the NSA, and other agencies, as required, supporting the specific needs of the JFC. The Joint Staff J-2 is the NIST program's executive agent and has delegated the NIST mission to the Deputy Directorate for Crisis Operations (J-2O). The J-2O manages daily operations and interagency coordination for all. A JTF or JSOTF intelligence staff may also request the deployment of a NIST. Typically, a NIST will not be tailored below the JTF level. Army special operations units attached to conventional units can access NISTs through the supported unit's intelligence staff. Army special operations units will frequently work with a NIST to deconflict sources.

3-50. The Joint Staff J-2 NMJIC is the focal point for all defense intelligence activities in support of joint operations. The NMJIC is comprised of regional analysts, target analysts, operational specialists, terrorism analysts, warning intelligence officers, and collection managers from the Joint Staff J-2. Additionally, the DIA has two elements collocated with the NMJIC—the defense collection coordination center and the MASINT operations coordination center. Additionally, elements of the NGA, the CIA, the NSA, representatives of the Services and, as required, other federal agencies are integral components of the NMJIC. Army special operations units typically access the NMJIC through JTF or JSOTF intelligence staffs.

3-51. Several other national agencies may have representatives at a NIST or may provide support liaison officers or teams to Army special operations units. These national agencies include—
- Department of Homeland Security.
- Department of State.
- Department of Justice.
- Department of Energy.

3-52. The list above is by no means exhaustive as the potential exists for virtually every federal department and agency to take part in joint DOD and civilian homeland security operations. Further information on intelligence support in interagency operations can be found in Appendix C.

Chapter 4
Intelligence Support to Civil Affairs

CA units are organized, trained, and equipped specifically to conduct CAO and support the commander in planning and conducting CMO. CA units provide the military commander with expertise on the civil component of the operational environment. The commander uses CA forces' capabilities to analyze and influence the population through specific processes and dedicated resources and personnel. As part of the commander's CMO, CA forces conduct operations nested within the overall mission and intent. CA forces help ensure the legitimacy and credibility of the mission by advising on how to best meet the moral and legal obligations to the people affected by military operations. The key to understanding the role of CA is recognizing the importance of leveraging each relationship between the command and every individual, group, and organization in the operational environment to achieve a desired effect. This chapter discusses the mission, organization, and intelligence requirements of CA units and CAO. It addresses the intelligence support provided to CA units, and the support CAO provide to the intelligence process.

MISSION

4-1. The mission of CA forces is to engage and influence the civil populace by planning, executing, and transitioning CAO in Army, joint, interagency, and multinational operations to support commanders in engaging the civil component of their operational environment, in order to enhance CMO or other stated U.S. objectives before, during, or after other military operations. Prior to employing CA forces, the distinction between active and passive gathering is stressed to commanders. CA Soldiers only gather information passively which helps establish and maintain their credibility with relevant populations.

4-2. CAO consist of the following core tasks:
- Populace and resources control.
- Foreign humanitarian assistance.
- Civil information management.
- Nation assistance.
- Support to civil administration.

4-3. CAO are those military operations planned, supported, executed, or transitioned by CA forces through or with the indigenous populations and institutions, international government organizations, nongovernmental organizations, or other government organizations to modify behaviors, to mitigate or defeat threats to civil society, and assist in establishing the capacity for deterring or defeating future civil threats in support of CMO or other U.S. objectives.

INTELLIGENCE REQUIREMENTS

4-4. Before deploying into any operational area—whether by friendly agreement, as part of a liberating force, or in an occupational role—CA units develop information requirements. The necessity to gather information on a specific area and its people, and on source material and agencies relevant to the operation, is essential to mission preparation and execution.

4-5. The CA functional specialists (found only in U.S. Army Reserve CA units) provide broad guidelines for CA information requirements that must be satisfied before and during deployment. It is imperative that requests for information are fed into the intelligence process as early as possible so unit intelligence personnel can either answer them based on existing databases or push them forward in the form of requests for information. Such information requirements might include, but are not limited to—

- Topography, hydrography, climate, weather, and terrain, including landforms, drainage, vegetation, and soils.
- Census, location, ethnic composition, and health factors of the population.
- Attitudes, beliefs, and behaviors of the population, including ideological, religious, and cultural aspects.
- Government structure, including forms, personalities, existing laws, and political heritage.
- Educational standards and facilities and important cultural activities and repositories.
- Communications, transportation, utility, power, and natural resources.
- Labor potential, including availability by type and skill, practices, and organizations.
- Economic development, including principal industries, scientific and technical capabilities, commercial processes, banking structure, monetary system, price and commodity controls, extent and nature of agricultural production, and accustomed population dietary habits.
- Cores of resistance movements.
- Organization and operation of guerrilla forces in rear areas and the extent and degree of volition involved in local support.
- Hostile activities, including espionage, sabotage, and other factors of subversion and disaffection.

INTELLIGENCE ORGANIZATION

4-6. The CA unit intelligence officer and his staff are the only military intelligence assets organic to the CA unit. The CA commander directs the intelligence process through his intelligence staff and the CMO center. Duties of the intelligence staff include the following:

- Supervises organic and attached intelligence assets.
- Integrates CA intelligence efforts with other units' and agencies' efforts and products.
- Establishes liaison with HN military and government agencies, as required.
- Coordinates with the chief of the security assistance organization and the area coordination center in each operational area to meet the commander's intelligence needs.
- Assesses enemy CA capabilities, potential COAs, and their effect.
- Integrates with other CA and supported-unit staffs.
- Produces and disseminates intelligence products and CA/CMO estimates.

4-7. Organic intelligence assets within CA units are relatively limited. Therefore, a key component of successful CA and/or CMO intelligence support is the ability of CA unit intelligence personnel to integrate themselves into the theater intelligence architecture through the intelligence staff of their higher HQ and supported units.

4-8. In garrison, CA personnel obtain, analyze, and record information in the advance of need. This base line information is compiled in the CA area study. Running estimates and assessments update this information as necessary. The primary compilation, analysis, and fusion of this information is done within the civil information cell within the civil-military operations center. The intelligence staff will use these documents to develop and maintain an information database for the unit's geopolitical focus. This database includes historical and geographical information, current periodic intelligence reports, and partially developed products for contingency or crisis planning. Potential CAO that support theater operational plans and contingency plans receive highest priority during the development process. Operational plans, operation orders, campaign plans, and supporting CA and intelligence annexes contain specific CA and/or CMO intelligence requirements. Necessary requirements are validated, prioritized, and incorporated into overall collection efforts.

Intelligence Support to Civil Affairs

NONORGANIC INTELLIGENCE SUPPORT

4-9. Since CA units have limited organic intelligence assets, nonorganic intelligence support is required. Nonorganic support enables the intelligence staff to develop accurate, complete, and timely intelligence to help the commander estimate the influence of CA factors on the mission or potential COAs.

INTELLIGENCE DISCIPLINES SUPPORTING CIVIL AFFAIRS

4-10. Intelligence support from the following disciplines is typically provided to CA by the supported unit, from the theater level downward:

- SIGINT assets, which include the SOT-As assigned to SF units and assets with longer-range capabilities assigned to conventional military intelligence units can be accessed through the intelligence staff of their parent unit. CA units use SIGINT to pinpoint telecommunications and mass media facilities in target areas and to help assess the effectiveness of CMO by monitoring hostile forces and relevant populations, groups, and institutions in the area.
- GEOINT and IMINT products, ranging from photos taken by a reconnaissance patrol to those taken by national assets, can be requested and obtained through intelligence staff channels. CA units can use IMINT to locate and determine the operational status of key civil infrastructure in denied areas. The GEOINT/IMINT used by CA units can include identifying and evaluating the operational capabilities of transportation networks, factories, and other public structures and systems. CA units, using their DOD activity address code, may requisition standard NGA products through the Army supply system directly from the Defense Logistics Agency, the appropriate installation map depot, or the supporting OCONUS Army map depot. USASOC helps its subordinate units obtain special GI&S products through the NGA liaison.
- TECHINT assets normally are assigned only at theater level and higher, but their products can be requested through intelligence staff channels and the theater JIC. CA units use TECHINT to identify key technical characteristics and specifications of construction equipment, industrial facilities, and utilities in target areas.
- OSINT assets, through publications, academics, and mass media, can provide information on natural disasters, biographic information, culture, historical context, weather, and even battle damage assessment. Open-source data can also be purchased for geospatial and mapping data. Less obvious is the use of open-source materials such as graffiti and tagging to identify gang turf or to gauge public opinion.
- HUMINT assets are available from the units normally assigned to a task force, and from the military intelligence battalions and brigades assigned to conventional units from division to theater level. CA units use HUMINT to help determine the extent of war damage in threat-controlled areas. HUMINT can also help to locate key technical personnel who can be of use in repairing or operating key infrastructure once friendly forces arrive.
- CI agents and analysts are available to CA units only through their supported units. CI Soldiers are trained to detect, evaluate, counteract, and prevent foreign intelligence collection, subversion, sabotage, and terrorism. CI support is essential to the security efforts of CMO.

NATIONAL AND THEATER SUPPORT

4-11. As is the case with all Army special operations unit theater intelligence activities, the CA unit intelligence staff coordinates with the intelligence staff of his supported unit for access to intelligence support. Likewise, CA intelligence staffs typically access national-level intelligence support through their supported unit. The theater JIC is the primary source for all-source intelligence support to all in-theater SOF.

4-12. With guidance from the CA unit commander, the intelligence staff prioritizes, validates, and consolidates all standing and routine requests for information that are being submitted to the JIC. To integrate and support CA intelligence requirements, higher HQ through the theater intelligence staff ensure that the JIC—

- Responds to CA/CMO-sponsored requests for information by integrating them into the theater requirements list.

Chapter 4

- Monitors requests for information status until the appropriate collection assets respond.
- Maintains an intelligence database to support requirements.

WEATHER SUPPORT

4-13. Weather support for CMO can be obtained through SOTF intelligence staffs, which have access to the United States Air Force (USAF) SOWTs. Regardless of the primary mission, CA units must have advanced knowledge of seasonal and irregular weather patterns. Direct weather support should include, but not be limited to—

- Weather advisories and warnings.
- Long-range weather forecasts.
- Precipitation patterns.
- Wind patterns.
- Tidal data.

4-14. Natural disasters are often the impetus behind CMO missions such as in the case of Hurricane Katrina. Other natural disasters, such as earthquakes, tornadoes, and floods, often require the deployment of U.S. forces to assist in disaster relief efforts within the United States and on foreign soil. An example is Operation UNIFIED RESPONSE, the 2010 Haiti earthquake relief operation. Severe weather during combat and noncombat missions may require the diversion of resources to relief operations.

CIVIL AFFAIRS SUPPORT TO THE INTELLIGENCE PROCESS

4-15. CA forces are chartered to target and engage the population in an overt manner consistent with U.S. policy and goals. CA forces project a nonthreatening posture in their mission, demeanor, and equipment; therefore, HN forces often afford CA personnel better access and placement with regard to people, places, and information. CA forces keep intelligence and intelligence-related activities separate from their operations in order to maintain the goodwill and freedom of maneuver they require to perform their mission. CA personnel are in an ideal position to passively collect a variety of information that may be of intelligence value, even though it is not their primary mission. They can be used to gain access and placement to select individuals and areas—wittingly or unwittingly—without sacrificing their mission. It is imperative that CA forces do not present the appearance of collecting intelligence.

4-16. In the routine preparation for and conduct of CAO, relevant information is entered into the intelligence process and becomes part of the finished IPB products. This information may include—

- *Demographics*: Shows dominant racial, religious, cultural, or political population densities. The intelligence staff uses these overlays to create templates of prevailing attitudes and loyalties in heterogeneous populations.
- *Infrastructure*: Depicts public utilities by showing the location and capability or capacity of all public utility nodes (such as power stations and substations, pumping stations, telephone company switches, and waste-handling facilities). These overlays, when used with maneuver overlays, can project the impact combat operations will have on the local population's ability to maintain basic living conditions.
- *Medical support*: Illustrates available health services support showing the location of private and public health service facilities (such as hospitals, pharmacies, and doctors, dentists, and veterinary offices). These overlays should reveal details such as capacity, age, capabilities, and equipment about each facility.
- *Protected or restricted targets*: Pinpoints locations of hospitals, national monuments, religious shrines or houses of worship, and other places protected by the Laws of War and the Geneva Convention.
- *Displaced civilians*: Shows population displacement. These overlays include—
    - Projected overlays showing the routes the displaced population most likely will use given a set of projected conditions (for example, disruption of the food supply or physical destruction of an urban area).

- Current situation overlays showing routes currently in use by the displaced population, including the refugee camps that have developed or are beginning to develop and the associated major supply routes affected by displaced civilians.

CIVIL AFFAIRS AREA STUDY AND AREA ASSESSMENT

4-17. The area study is a process common to all CAO. Area study files contain information on a designated area. This information supports contingency and special operations planning in areas assigned. CA personnel obtain, analyze, and record information in advance of need. They update the study as required through an area assessment. An area study has no single format. The information acquired through the area study supports the area assessment. An area assessment begins with receipt of the mission. CA area assessments that support other forces should supplement—not repeat—information in the basic area study. Area studies provide intelligence staffs valuable input on demographics, infrastructure, displaced civilians, and other civil areas. An area assessment is done after the unit arrives and will confirm the area study, or reveal changes or modifications as necessary for the situation. Both the area study and assessment may provide in-depth information to satisfy intelligence requirements for both special operations and conventional forces. FM 3-57 includes more information on area studies and assessments.

ASCOPE ANALYSIS

4-18. CAO/CMO planners use the operational variables of PMESII-PT and the mission variables of mission, enemy, terrain and weather, troops and support available, time available, and civil considerations (METT-TC) to analyze, develop, and establish COAs during the MDMP. Civil considerations are analyzed using the mnemonic ASCOPE. The information resultant from ASCOPE analysis can be invaluable to intelligence staffs when properly integrated through the intelligence process. The six characteristics are discussed in the following paragraphs.

Areas

4-19. Areas are key localities or aspects of the terrain within a commander's operational environment that are not normally thought of as militarily significant. Failure to consider key civil areas, however, can seriously affect the success of any military mission. CAO planners analyze key civil areas from two perspectives: how do these areas affect the military mission and how do military operations impact on civilian activities in these areas? At times, the answers to these questions may dramatically influence major portions of the COAs being considered.

Structures

4-20. Existing civil structures take on many significant roles. Some, such as bridges, communications towers, power plants, and dams, are traditional high-priority targets. Others, such as churches, mosques, national libraries, and hospitals, are cultural sites that are generally protected by international law or other agreements. Still others are facilities with practical applications, such as jails, warehouses, schools, television stations, radio stations, and printing plants, which may be useful for military purposes. Structures analysis involves determining the location, functions, capabilities, and application in support of military operations. It also involves weighing the consequences of removing them from civilian use in terms of political, economic, religious, social, and informational implications; the reaction of the populace; and replacement costs.

Capabilities

4-21. Civil capabilities can be viewed from several perspectives. The term capabilities may refer to—
- Existing capabilities of the populace to sustain itself, such as through public administration, public safety, emergency services, and food and agriculture systems.
- Capabilities with which the populace needs assistance, such as public works and utilities, public health, public transportation, economics, and commerce.

- Resources and services that can be contracted to support the military mission, such as interpreters, laundry services, construction materials, and equipment. Local vendors, the HN, or other nations may provide these resources and services. In hostile territory, civil capabilities include resources that may be taken and used by military forces consistent with international law.

4-22. Analysis of the existing capabilities of the operational area is normally conducted based on the CA functional specialties. The analysis also identifies the capabilities of partner countries and organizations involved in the operation. In doing so, CAO/CMO planners consider how to address shortfalls, as well as how to capitalize on strengths in capabilities.

Organizations

4-23. Civil organizations are organized groups that may or may not be affiliated with government agencies. They can be church groups, fraternal organizations, patriotic or service organizations, and community watch groups. They might be international government organizations or the nongovernmental organization community. Organizations can assist the commander in keeping the populace informed of ongoing and future activities in an operational area and influencing the actions of the populace. They can also form the nucleus of humanitarian assistance programs, interim-governing bodies, civil defense efforts, and other activities.

People

4-24. People, both individually and collectively, have a positive, negative, or neutral impact on military operations. In the context of ASCOPE, the term people include civilians or nonmilitary personnel encountered in an operational area. The term may also extend to those outside the operational area whose actions, opinions, or political influence can affect the military mission. In all military operations, U.S. forces must be prepared to encounter and work closely with civilians of all types. When analyzing people, CA Soldiers consider historical, cultural, ethnic, political, economic, and humanitarian factors. They also identify the key communicators and the formal and informal processes used to influence people.

4-25. Regardless of the nature of the operation, military forces will usually encounter various civilians living and operating in and around the supported unit's operational area. To facilitate determining who they might be, it is useful to separate civilians into distinct categories. In foreign operations, these categories might include—

- Local nationals (town and city dwellers, farmers and other rural dwellers, and nomads).
- Local civil authorities (elected and traditional leaders at all levels of government).
- Expatriates.
- Foreign employees of international government organizations or nongovernmental organizations.
- U.S. Government and third-nation government agency representatives.
- Contractors (U.S. citizens, local nationals, and third-nation citizens providing contract services).
- DOD civilian employees.
- The media (journalists from print, radio, and visual media).

4-26. Civilian activities are dictated primarily by the type of environment in which they occur. Each category of civilian should be considered separately, as their activities will impact differently, both positively and negatively, on the unit's mission. Military operations affect civilian activities in various ways. Commanders should consider the political, economic, psychological, environmental, and legal impact of operations on the categories of civilians identified in the operational area.

Events

4-27. As there are many different categories of civilians, there are many categories of civilian events that may affect the military mission. Some examples are planting and harvest seasons, elections, riots, and evacuations (both voluntary and involuntary). Likewise, there are military events that impact the lives of civilians in an operational area. Some examples are combat operations, including indirect fires, deployments, and redeployments. CAO/CMO planners determine what events are occurring and analyze the events for their political, economic, psychological, environmental, and legal implications.

CIVIL INFORMATION MANAGEMENT

4-28. Civil information is information developed from data with relation to civil areas, structures, capabilities, organizations, people, and events, within the civil component of the commander's operational environment that can be fused or processed to increase DOD/interagency/international government organization/nongovernmental organization/indigenous populations and institutions situational awareness, situational understanding, or situational dominance. Civil information management is the process whereby civil information is collected, entered into a central database, and internally fused with the supported element, higher HQ, interagency, international government organizations, and nongovernmental organizations to ensure the timely availability of information for analysis and the widest possible dissemination of the raw and analyzed civil information to military and nonmilitary partners throughout the operational area. It is essential that relevant intelligence staffs are aware of the information available through civil information management. Three essential elements of civil information management are—

- *Civil reconnaissance.* Civil reconnaissance is a targeted, planned, and coordinated observation and evaluation of those specific civil aspects of the environment. Civil reconnaissance focuses specifically on the civil component through ASCOPE analysis. Civil reconnaissance can be conducted by CA forces or by other forces, as required. Examples of other specialties and assets that can conduct civil reconnaissance are MISO, engineers, medical personnel, military police, and unmanned aircraft systems (UASs).
- *Civil information grid.* The civil information grid provides the capability to coordinate, collaborate, and communicate to develop the civil components of the common operational picture. The civil information grid increases the situational understanding for the supported commander by vertically and horizontally integrating the technical lines of communication. This framework links every CA Soldier as a sensor and consumer to the civil information management cell of the CMO center and the CMO cell.
- *Supporting tasks.* Generally, CA Soldier civil information management tasks include—
  - Conducting civil reconnaissance to find, analyze, and report civil information.
  - Coordinating with non-CA assets to achieve a coherent reconnaissance and execution plan.
  - Synchronizing the collection and consolidation of civil information.
  - Developing the civil components of the common operational picture.
  - Increasing the supported commander's environment awareness.
  - Assisting in the development of the supported commander's common operational picture.
  - Conducting interagency, international government organization, nongovernmental organization, and indigenous populations and institutions coordination.

This page intentionally left blank.

Chapter 5
Intelligence Support to Military Information Support Operations

MISO are a flexible and vital part of any graduated response to international crisis or conflict and provide means to support GCCs' engagement strategies during peacetime. The Military Information Support Operations Command (MISOC) supports DOD and other government agency operations by planning, developing, and executing MISO across the range of military operations. Intelligence support to MISO frequently focuses on unique information requirements. Early integration of intelligence requirements into the supported unit's collection plan is critical. Conversely, failure by supported units to recognize the unique abilities of trained personnel to gather information relevant to their areas of expertise may result in duplication of effort or outright failure to accurately assess the perceptions and vulnerabilities of an adversary. Specific intelligence requirements to MISO are covered in this chapter.

MISSION

5-1. MISO are planned operations to convey selected information and indicators to foreign target audiences to influence their emotions, motives, objective reasoning, and ultimately the behavior of foreign governments, organizations, groups, and individuals. The purpose of MISO is to induce and reinforce foreign attributes and behavior favorable to the originator's objective. MISO include planned series of products and actions that are directed toward specified target audiences. As the primary information-related capability for influence employed by information operations, MISO support decisive action throughout the range of military operations.

5-2. The MISOC and its subordinate elements (Military Information Support groups [MISGs]) execute the following mission components:
- Rapidly deploy assigned forces to support Army conventional or special operations forces.
- Develop supporting MISO plans and integrate them with operational level plans and theater peacetime MISO programs.
- Conceive, develop, and produce MISO messages and psychological actions (PSYACTs).
- Conduct operational-level MISO by establishing a joint MISO task force HQ, directing and employing multi-Service MISG assets, and liaising with Service components, host nation, and multinational partners to facilitate execution.
- Conduct MISO in concert with and in support of ground operations.
- Provide linguists with cultural experience to assist commanders executing MISO.
- Prepare basic and special MISO intelligence assessments and studies for the Chairman of the Joint Chiefs of Staff, unified commands, and other government agencies as directed by USSOCOM.

INTELLIGENCE REQUIREMENTS

5-3. Accurate, timely, and relevant intelligence and information are required for planning, executing, and evaluating the effectiveness of MISO messages and actions. Continual and updated intelligence about the operational environment and its local populations facilitate the decisionmaking process for commanders

Chapter 5

and MISO planners to target the most appropriate audiences with the most appropriate messages. This necessity for detailed intelligence drives the need for access to theater and national systems.

5-4. The requirements for MISO are extensive and are consistent with those of the supported unit, such as current intelligence, background studies, intelligence estimates, and cultural information. MISO forces require intelligence support to analyze populations and their environment, and to evaluate the effectiveness of messages and actions on them.

5-5. MISO activities require continuous and timely intelligence to assess initial target audience baseline attitudes, perceptions, and behaviors, and any subsequent changes in target audience decisionmaking and actions derived after the application of MISO. MISO forces rely on a number of sources of information to understand the history, geography and terrain, culture, political, economic, social, and military factors of a country or region. Much of this information is obtained from a variety of open-source resources. Reconnaissance and surveillance assets are required for detailed, verifiable facts and data, and analyses that are unavailable through unclassified sources. Examples of this required information include—

- Evaluations of threat capabilities, vulnerabilities, and probable COAs.
- The potential for adversary messages and actions.
- Accessibility of potential target audiences.
- Population motivations and current behaviors.
- Indicators of effectiveness of MISO messages on intended and unintended audiences.
- Populace reactions to friendly, hostile, and neutral force actions and messages.

5-6. Target audience information includes the identity, location, conditions, vulnerabilities, susceptibilities, and effectiveness of a specified target audience. During deliberate or crisis action planning, the intelligence staff accesses existing databases and products to support the mission. They query every available source, including the TSOC intelligence staff, the theater JIC, and national, HN, and the supported units' resources. The intelligence staff also refers to current versions of special MISO studies and special MISO assessments pertaining to the target audience. These resources are processed and integrated as part of MISO planning and target audience analysis.

5-7. Measures of effectiveness for MISO are observable and measurable impact indicators that help determine the degree of effectiveness MISO have been in achieving specified objectives. For some measures of effectiveness, the unit may be able to collect complete information that is easily validated by assigned personnel; for example, the percentage of a target audience voting in an election may be verified through reliable media sources or international government organizations. Measures of effectiveness requiring intelligence support are submitted as information requirements which are subsequently nominated for inclusion in the intelligence collection plan. Intelligence staffs can liken MISO impact indicators to the battle damage assessment factors of lethal munitions. Some example impact indicators include—

- Number of defectors from a unit or organization.
- Number of attacks against U.S. or HN forces.
- Amount of weapons turned in to an amnesty or buyback program.
- Percentage of population voting.

5-8. The MISGs provide capabilities to support the countering of adversary information activities. Whether or not the MISGs or their subordinate units are supporting a specific COA to counter adversary information activities, all of its employed elements must be aware of all information activities conducted within their area of responsibility and area of interest. Intelligence staffs keep in mind that the area of interest for MISO may be regional, transregional, or global. MISO units need to know adversary or enemy information capabilities. To provide intelligence support, all intelligence disciplines are used in varying degrees. Scientific and technical intelligence plays a vital role in revealing frequencies, video formats, print capabilities, and Internet capabilities. Technical factors are necessary to determine the sophistication of the producer of the information product and may reveal the identity of the producer. While other disciplines such as OSINT and SIGINT may provide some information about adversary information activities, CI assets may be the only discipline to provide specific required information about adversary information activities.

5-9. Information requirements about adversary propaganda and information activities center around—
- *Source*: The originator and creator of the product.
- *Content*: A thorough analysis of the message. Intelligence support to content analysis focuses on deriving the correct translation of the language used and determining the meaning of previously unknown symbols or images.
- *Audience*: Intelligence support to help determine the four classes of audiences associated with the product—apparent, intermediate, unintended, and ultimate.
- *Media*: The technical details and specifications of the media used in the product. Multiple intelligence disciplines may play a part in determining these details.
- *Effects*: The effect of the product on which it was intended. HUMINT and OSINT contribute significantly in determining the effects of information activities and products.

5-10. Detainees can provide vital information to MISO efforts. Intelligence obtained from detainees can assist in determining the effectiveness of MISO series and provide critical target audience analysis updates. MISO units tasked with supporting detainee holding areas may be able to interview detainees to determine the effectiveness of MISO series and to obtain feedback on product concepts and prototypes. It is unlikely that MISO units will provide direct support to all detainee holding areas and therefore they rely on intelligence staffs to collect intelligence on detainee demographics, detainee exposure to MISO products, and the effects of those products on the detainees. Whether collected by MISO units in detainee holding areas or obtained through intelligence staff support, MISO units rely on organic and nonorganic military intelligence Soldiers to apply the intelligence process to all information collected from detainees and transform it into reliable, timely intelligence that supports MISO planning, developing, message delivery, and assessment.

INTELLIGENCE ORGANIZATION

5-11. Organic MISO unit intelligence assets include military intelligence Soldiers, MISO specialists, and DA civilian analysts. MISO specialists and civilian analysts augment military intelligence personnel in creating a unified collection effort. Psychological Operations (PSYOP) Soldiers and civilian analysts gather information and collate already processed intelligence. Unlike military intelligence personnel and trained CI personnel, they are not active intelligence collectors.

RESEARCH AND ANALYSIS

5-12. The MIS battalions have Army civilian analysts assigned to the five cultural intelligence sections. These analysts have extensive regional, cultural, and language expertise and focus on the psychological and demographic information and intelligence specific to the mission of changing foreign target audience behavior. The analysts conduct original research using open sources as well as various intelligence databases to produce special MISO studies, special MISO assessments, and other finished intelligence reports. Each cultural intelligence section supports a regionally oriented MIS battalion. The MISOC is responsible for the MISO portions of the Defense Intelligence Analysis Program. Studies written by the cultural intelligence section analysts support that responsibility.

5-13. Analysts in the cultural intelligence sections conduct research and analysis of target countries, regions, groups, and issues to support MISO development. The detachment provides timely political, cultural, social, political-military, economic, and policy analyses to the MISOC commander, subordinate units and staffs, geographic combatant command MISO planners, and other agencies. The analysts assist in deliberate and contingency planning as well as execution of approved programs. The studies and other internally generated analytical products are accessible through the unit's home page on INTELINK (SCI) and INTELINK (Secret). The Psychological Operations Automated System Message Center provides the MISO community with access to classified and unclassified databases.

Chapter 5

GROUP INTELLIGENCE STAFF

5-14. The intelligence staff of the MISG facilitates the planning for collection and dissemination of intelligence required to conduct MISO. When the unit's battalions, companies, and detachments operate autonomously or in support of other forces, the senior intelligence officer is usually the battalion's intelligence staff officer. When the unit's intelligence officer is unavailable, the detachment commander establishes the same relationship with the supported unit's intelligence officer as that normally established between a supporting unit's intelligence officer and the supported unit's intelligence officer.

MILITARY INFORMATION SUPPORT BATTALION

5-15. The MIS battalion intelligence staff performs the same intelligence tasks as the group intelligence staff. Typically, a MIS battalion forms the core of a MIS task force or joint MIS task force when established for a major operation. In these instances, the battalion intelligence staff forms the nucleus of the MIS task force intelligence staff.

5-16. All Regular Army ARSOF MIS battalions are subordinate to a group and are organized and equipped to provide dedicated support to a specific GCC. Each battalion operates as a modular unit at the tactical and operational levels capable of planning, developing, delivering, and assessing MISO with assigned unit and communication equipment. When communication and development requirements exceed the capacity of battalion assets, the group's organic dissemination battalion provides an additional capability. This battalion maintains organic regional target analysis, production, distribution, and dissemination assets. Each battalion has an organic cultural intelligence section with regional expertise. Personnel from the MIS battalion conduct extensive open-source research on specified target audiences and maintain regional and country knowledge bases and systems. This information is validated through intelligence support organizations and structures. MISO units monitor, collate, and analyze adversary information activities and messages to support the potential countering of their effects. MISO forces conduct surveys and interviews to support their operations. These activities should not be confused with tactical intelligence collection.

NONORGANIC INTELLIGENCE SUPPORT

5-17. Several agencies and organizations habitually provide intelligence support to MISO. Typically, MISO units access this support through their supported unit. The following paragraphs cover how the various intelligence disciplines support MISO and specific agencies, organizations, and units that may support MISO.

INTELLIGENCE DISCIPLINES SUPPORTING MILITARY INFORMATION SUPPORT OPERATIONS

5-18. MISO personnel use data from all of the intelligence disciplines to plan their missions. This requires intelligence staffs to acquire and use information from national and theater organizations. Information requirements for MISO units are often passed from the supported unit to higher echelons. Specific ways nonorganic assets in all intelligence disciplines support MISO are discussed in the following paragraphs.

Human Intelligence Support

5-19. Intelligence and information gathered from EPWs, defectors, line crossers, refugees, captured documents, and published materials often provide MISO units with significant insights into the psychological situation in a specific area or within a target audience. With consent and proper authority, these sources may be used to develop and test MISO products.

5-20. HUMINT support for target audience assessments is requested from the supported unit and theater intelligence assets. National and other special operations units' capabilities and resources may be requested as well. HUMINT collection trained personnel are located at all EPW collection points and holding facilities at echelons division and above. MISO units require information that only HUMINT collectors can provide; they must coordinate their requirements with the command or organization that has the collection capability. Operating forces below division level with assigned or attached human collection teams may be a source of support particularly for tactical MISO units.

Signals Intelligence Support

5-21. SIGINT assets support MISO units by providing products extracted from locating, monitoring, and transcribing adversary communications. SIGINT assets provide information on the broadcast footprints of radio and television transmitters that deliver adversary information, and deliver neutral or friendly information. These assets provide information and intelligence that help reveal enemy activities or plans so that MISO forces can develop effective countermeasures.

Geospatial Intelligence Support

5-22. MISO units request imagery analysis support from the supported command. MISO personnel use imagery analysis in various ways. Imagery analysis helps locate and determine the capabilities and operational status of media production and dissemination assets. Personnel use imagery to locate mobile target audiences. By analyzing structures, dress, local art, or even graffiti, PSYOP Soldiers can determine the ethnicity, religions, and other demographic data of a population center. Other uses for imagery analysis products include identifying and evaluating operational capabilities of transportation networks, factories, and other public structures or systems. Such analysis of infrastructure can confirm or deny suspected vulnerabilities targetable by a MISO series. Imagery analysis can confirm or deny acts of rioting, acts of sabotage, demonstrations, work slowdown, and other public gatherings that are impact indicators for a program, series, or specific product.

Open-Source Intelligence Support

5-23. OSINT obtained through publications, academic institutions, and mass media can provide information on natural disasters, biographic information, culture, historical context, weather, and battle damage. All of the political/military factors can be researched in large part through OSINT. The most powerful OSINT tool is the Internet which must be used with caution as many sites contain no way to validate the source of their information. Open-source materials range to less-obvious sources such as graffiti to identify gang turf, subcultures, or social groups, or T-shirts and bumper stickers that may help gauge public opinion.

Technical Intelligence Support

5-24. MISO units use TECHINT to focus their efforts on critical technical adversary units, entities, and installations. They identify alternative methods of message dissemination through the analysis of the target population's information infrastructure. The captured materiel exploitation center or a battlefield TECHINT team at corps produces TECHINT products. Specific requests for TECHINT support are coordinated through intelligence staffs.

Counterintelligence Support

5-25. CI detects, evaluates, counteracts, or prevents foreign intelligence collection, subversion, sabotage, and terrorism. It determines security vulnerabilities and recommends countermeasures. CI operations support operational security, deception, information protection, operational area security, and countering adversary information activities. MISO units request CI support through the supported unit.

OTHER NONORGANIC SUPPORT

5-26. MISO units plan, produce, and disseminate the series required to accomplish their missions. MISO specialists and organic assets obtain, process, and use considerable volumes of information on specified target audiences necessary to execute successful MISO. Target audiences frequently include groups at high levels of foreign governments and organizations. The potential psychological impacts of friendly actions may have implications on transregional or global scales. Organic intelligence Soldiers may be unable to provide enough support for mission success. Therefore, MISO unit staffs must leverage the support available in the intelligence infrastructure to obtain needed intelligence and information.

National Support

5-27. The majority of MISO, particularly at the strategic and operational levels, require access to intelligence information and products produced at the national level. Other government agencies, such as the CIA and Department of State, collect and produce intelligence valuable to MISO. These agencies are engaged in every region of the world and are sanctioned to provide support for MISO intelligence purposes. The NMJIC, the NGIC, and the NSA are excellent sources for intelligence reports and products. These agencies have extensive knowledge of potential target audiences. They also have databases and collection frameworks in place with developed intelligence requirements and tasks that can support MISO efforts. MISO units can obtain relevant intelligence and information through the joint intelligence operations center at USSOCOM.

Theater Support

5-28. The GCC's intelligence staff forms theater intelligence policy and plans, and coordinates intelligence support for deployed SOF, including MISO units. The intelligence staff—

- Ensures that sufficient intelligence support is available for each mission tasked by the commander.
- Relies on the theater Service intelligence organizations to collect, produce, and disseminate intelligence to meet MISO requirements.
- Tasks subordinate units and coordinates with national agencies to collect and report information in support of MISO intelligence requirements.

5-29. Theater operation orders, operation plans, campaign plans, and supporting MISO and intelligence tabs and annexes contain specific MISO information requirements. Most of these requirements are validated and incorporated into MISO and intelligence collection plans. To meet some of these requirements, senior intelligence officers may need to reinforce or refocus available assets. Often, the senior intelligence officer must access information or intelligence from other units, intelligence agencies, or sources at higher, lower, and adjacent echelons. Figure 5-1 shows examples of theater sources that support MISO.

Units and Agencies	Agencies Outside the Intelligence Community	Other Contributors
HN/Partner Nation (PN) Military	HN/PN Government Agencies	EPWs
HN/PN Paramilitary	HN/PN Police	Detainees
SF	Department of State	Civilians
CA	Department of Justice	Nongovernmental Organizations
Military Intelligence	Department of Treasury	Media Broadcasts
Military Police	Department of Commerce	Printed Material
CIA	Drug Enforcement Agency	Internet
DIA		
NMJIC		
NGIC		
NGA		
Open Source Center		
NATO		
Allied Military Counterparts		

Figure 5-1. Potential intelligence sources for Military Information Support Operations

Tactical Support

5-30. MISO unit intelligence personnel must coordinate with higher, adjacent, and supported units for access to their tactical HUMINT. Swift exploitation of collected HUMINT is critical. Intelligence staffs supporting tactical MISO must ensure there are enough conduits to support the exchange of perishable information to and from deployed MISO units.

Weather Support

5-31. Weather and other environmental factors affect almost all MISO dissemination missions. Severe weather may degrade MISO dissemination efforts, as in the case of airborne leaflet drops. Sunspot activity can disrupt radio and television broadcasts into a target area. Severe weather may also enhance MISO if it affects adversary morale. Therefore, MISO units need accurate weather and environmental information. Required weather support includes—
- Forecasts of general weather conditions and specific elements of meteorological data, as described in the 24-hour forecast.
- Solar, geophysical information, and climatic studies and analysis.
- Weather advisories, warnings, and specialized weather products, as required.

5-32. The primary source for required weather intelligence support, to include specialized products, is the USAF 10th Combat Weather Squadron. This squadron is a component of the Air Force Special Operations Command that provides SOWTs for attachment to Army special operations units.

GEOSPATIAL INFORMATION AND SERVICES AND OTHER INTELLIGENCE PRODUCTS

5-33. MISO units may requisition standard NGA products through the Army supply system directly from the Defense Logistics Agency. Intelligence products and services may be requested from the DIA. USASOC helps units obtain special GI&S through the NGA liaison.

MILITARY INFORMATION SUPPORT OPERATIONS SUPPORT TO THE INTELLIGENCE PROCESS

5-34. MISO units produce specialized analyses to support a variety of combat and intelligence missions and operations. MISO units develop these products by monitoring and assessing situations and evaluating the impact on specified target audiences and supporting MISO objectives. The main focus of this analytical effort is on diplomatic, information, military, and economic factors, with predominate emphasis on information factors. Analysis by MISO units examines the psychological impact of all actions taken to exercise these four elements of national power—diplomatic, information, military, and economic—in preventing or resolving conflict or in the prosecution of war. Products of MISO analysis include, but are not limited to—
- Strategic and operational documents such as special MISO studies and assessments.
- MISO reports and estimates.
- Target audience analyses.

5-35. Although MISO units primarily use these products to conduct their operations, the products also contain information and previously produced intelligence that is useful to SOF, other Army operating forces, and joint and national agencies. These products contain diverse information on social customs, enemy morale, key communicators, centers of gravity, and key nodes.

MILITARY INFORMATION SUPPORT OPERATIONS ASSESSMENTS AND ANALYSIS

5-36. In the course of conducting target audience analysis and media and local area assessments, MISO units gather diverse information from open sources. This information, including demographics and infrastructure, frequently has value to conventional and special operations units when processed into intelligence. MISO series development may yield information needed at theater or national levels. In stability operations, the commander's priority information requirements and critical information requirements often include requirements for information on the attitudes and perceptions of select groups within the area of responsibility. Whether or not these groups are designated as target audiences, MISO units may be tasked to collect these information requirements.

5-37. Tactical MIS teams have direct interactions with people that facilitate opportunities and the ability to collect information that is not afforded to all units. These teams disseminate MISO products, conduct face-to-face communications with target audiences, and gather information on specified target audiences and media resources through assessments. Coupled with a team's language capability (organic and/or attached or contracted interpreter support), these activities are an important passive collection resource that provides supported higher HQ and other MISO units with critical information on target audience attitudes and behaviors.

TARGET AUDIENCE ANALYSIS PROCESS

5-38. The target audience analysis process is a systematic and continuous process that MISO units use to plan and conduct series development. It is similar to the intelligence process, as it is continuous and relies on intelligence data presented in a format similar to IPB products. The target audience analysis process builds on existing IPB but is oriented to populations. The process examines target audiences within and outside of the targeted area of operations. The target audience analysis process examines—

- Attitudes, behaviors, conditions, and vulnerabilities of target audiences.
- Political analysis.
- Military analysis.
- Demographics.
- Geography and infrastructure.
- Media and broadcast frequency analysis.

These factors are analyzed for expected target audience behavior and to identify means and requirements for desired behavior changes.

5-39. Other analysis and information-gathering includes information collected from EPWs and detainees. This information can include detailed analysis of the psychological condition of detainees and the effects of both MISO and other military operations on targeted and nontargeted personnel. Interviews by MISO unit personnel are typically conducted after military intelligence units have screened and interviewed detainees. Further MISO interviews may yield updated information or may authenticate or rebut the accuracy of previous interviews. MISO unit analysis of adversary information activities, to include propaganda, may yield information valuable to other units, such as the Public Affairs Office or CA units. Analysis of threat and adversary propaganda and information capabilities and trends may yield indications and warnings of threat and adversary intentions or COAs.

ASYMMETRIC SOFTWARE KIT

5-40. The ASK serves as both a means of reporting information and of effectively using information gathered by PSYOP Soldiers. The ASK assists the user to develop and manage the MISO common operational picture. Critical to analysis and information-gathering is data-mining. Data-mining is performed using OrionMagic and Pathfinder. OrionMagic has a built-in Web browser and allows the user to organize information collected in an electronic filing system in the form of cabinets, outlines, folders, and note cards. Pathfinder for the Web is an intelligence data-mining and visualization application developed to assist analysts in the processing, searching, and analyzing of large quantities of text data.

5-41. The analysis and visualization are conducted using Analyst's Notebook and ArcGIS. Analyst's Notebook is used for link analysis. The targeting process can be aided and refined through link analysis using Analyst's Notebook. This visualization of analysis easily identifies links, patterns, and connections that should explain relationships and correlations between people, incidents, associated groups, and other useful information that can be associated. Further visualization can be conducted using ArcGIS. All requisite information can be represented geospatially using various programs in the ArcGIS suite. ArcGIS enables spatial information created to be queried as an aid in planning and targeting.

5-42. Within the MISO unit, an element is responsible for the reporting and management of the ASK. Which element is tasked with the responsibility depends on the task organization of the unit. Within bandwidth constraints, information from the ASK can be shared digitally. Information created in response to the commander's critical information requirements and priority information requirements, when created as shape files, will look identical from the reporting agency to the receiving agency. Tactical MIS detachments collect information in response to the commander's critical information requirements and priority information requirements through their subordinate tactical MIS teams and create the MISO common operational picture for their area of responsibility. Tactical MIS teams use their electronic news gathering kits to add important visual information to the common operational picture. This common operational picture is shared and reported up through the MISO coordination channel. Elements within the channel compile, analyze, and create the larger common operational picture, refining targeting, target audiences, and placement of MISO assets. The cultural intelligence sections and regional MIS battalions use the ASK for analytical purposes. Upon redeployment, units are debriefed and share information with the cultural intelligence section. The cultural intelligence section serves as the central repository for regionally oriented information within MISO units.

INFORM AND INFLUENCE ACTIVITIES

5-43. MISO are an information-related capability along with military deception, electronic warfare, operational security, and computer network operations. MISO facilitate the effectiveness of the other information-related capabilities by reinforcing desired adversary perceptions of a given situation. MISO establish a link between influence and behavior which distinguishes it from other information-related capabilities and activities. Information gathered by, and intelligence required by, MISO units is used within the information operations staffs to develop targeting and finalize target nominations. MISO personnel within information operations staffs may be tasked with collecting the commander's critical information requirements and priority information requirements that support information operations.

This page intentionally left blank.

Chapter 6
Intelligence Support to Rangers

Rangers provide a responsive strike force for conducting direct action missions normally as part of a SOTF. Ranger operations rely on the elements of surprise, precise planning, and orchestrated execution to conduct special missions supporting vital U.S. interests. There is one active Ranger Regiment with four Ranger battalions. Intelligence support for Ranger operations primarily focuses on providing target-specific information for objectives of strategic or national importance. At the three Ranger line battalions, the intelligence structure parallels that of a light infantry battalion, but is more robust. The organic intelligence structure within the Ranger regimental HQ is larger than that of an infantry brigade and is structured to provide linkage from higher echelons to the operational units. In addition, the Ranger Regiment has organic reconnaissance capacity in the form of the Ranger reconnaissance company.

MISSION

6-1. The Ranger Regiment's task is to plan and conduct direct action against strategic or operational targets in support of national or theater objectives. The Ranger Regiment and its subordinate battalions have a worldwide focus. Ranger missions are diverse and carried out on any terrain and under any condition. Ranger operations have high-risk and high-payoff attributes. Therefore, accurate, detailed, and timely intelligence is critical for planning and execution. It is through active interface with the supporting intelligence system that Ranger units receive answers to specific target information requirements since the Ranger Regiment has limited organic intelligence collection assets.

6-2. Ranger direct action operations may support or be supported by other special operations activities, or they may be conducted independently or with conventional military operations. Ranger direct action operations typically include—
- Raids against targets of strategic or operational value.
- Lodgment operations such as airfield seizures.
- Noncombatant evacuation operations.

INTELLIGENCE REQUIREMENTS

6-3. Intelligence support requirements for the Ranger Regiment center on target intelligence. Ranger forces conducting direct action are employed on targets of strategic or operational value. These targets may be similar in nature and location to targets being interdicted by conventional means—for example, a weapons of mass destruction facility—but because of the type of unit being employed, a higher level of intelligence resolution is required. For instance, a weapons of mass destruction facility being targeted for interdiction through the air tasking order may require using only SIGINT, MASINT, and GEOINT. A target being interdicted by a Ranger unit may require products from these disciplines as well as extensive use of HUMINT to bring the level of detail from the general exterior of the facility and a few salient physical features to a detailed product encompassing interior details, guard force composition, schedules, protocols, and mission command infrastructure. During a noncombatant evacuation operation, commanders must possess a detailed picture of civil considerations. Intelligence preparation for a noncombatant evacuation operation will typically require detailed data and intelligence support from national or joint agencies.

Chapter 6

INTELLIGENCE ORGANIZATION

6-4. Intelligence assets organic to the Ranger Regiment are organized according to operational and analytical needs. The Ranger Regiment has organic assets it can use to perform intelligence functions and missions. The regimental intelligence staff and the Regimental Military Intelligence Company are structured to support the—

- SOTF—a robust, sustained mission command and operational capability.
- Strike Force—short-duration capability in support of national missions.
- Home Station Operations Center—a sustained garrison mission command capability.

REGIMENTAL INTELLIGENCE STAFF

6-5. The regimental intelligence staff section is depicted in Figure 6-1. As the primary intelligence advisor to the commander, the intelligence staff officer—

- Oversees all intelligence operations for the 75th Ranger Regiment.
- Provides the commander with all-source intelligence assessments and estimates at the tactical, operational, and strategic levels.
- Directs tasking of intelligence collection assets, manages interrogation operations, interprets imagery from overhead and other systems, directs CI and operational security operations, and manages SIGINT operations, including jamming and participation in deception operations.
- Identifies, confirms, and coordinates area requirements for geospatial information and service products to support operational plans and contingency plans.

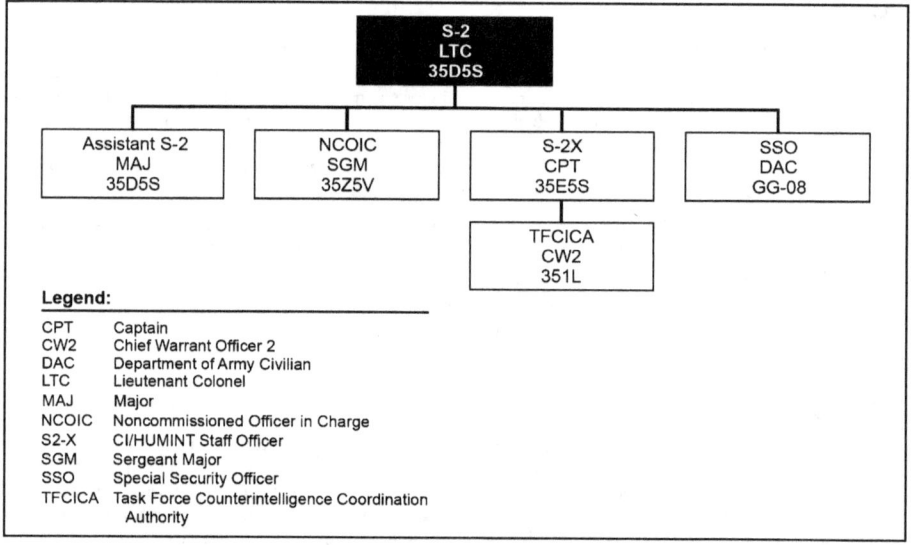

Figure 6-1. Regimental S-2

6-6. The assistant intelligence officer assists the intelligence officer with the aforementioned tasks and functions. The assistant intelligence officer is the senior military intelligence representative in the Regimental Alternate Assault Command Post. Additionally, he serves as the intelligence liaison to other special operations units, as required.

Intelligence Support to Rangers

6-7. CI and HUMINT operations are managed by the S-2X and the Task Force Counterintelligence Coordination Authority warrant officer. They are responsible for directing, coordinating, and advising on all CI and HUMINT operations within the regiment's area of responsibility. The S-2X and the Task Force Counterintelligence Coordination Authority warrant officer supervise the execution of the regimental CI and HUMINT collection plan, to include liaison and coordination with adjacent units in support of current operations. The S-2X interfaces with higher HQ and outside agencies to ensure adequate CI and HUMINT support to the regiment. The special security officer ensures that SCI material is properly secured, maintained, and accredited.

RANGER MILITARY INTELLIGENCE COMPANY

6-8. The Ranger Military Intelligence Company (Figure 6-2) is the principal source of intelligence support to Ranger commanders. The Ranger Military Intelligence Company's capabilities include the following:
- Sustained multidisciplined intelligence support to the SOTF.
- Ability to provide ground-based SIGINT operations.
- Sustained CI in support of SOTF operations.
- Sustained HUMINT operations in support of SOTF operations.
- Ability to interrogate EPWs.
- Detailed imagery, geospatial, and terrain products.
- Analytical reachback from the Home Station Operations Center.

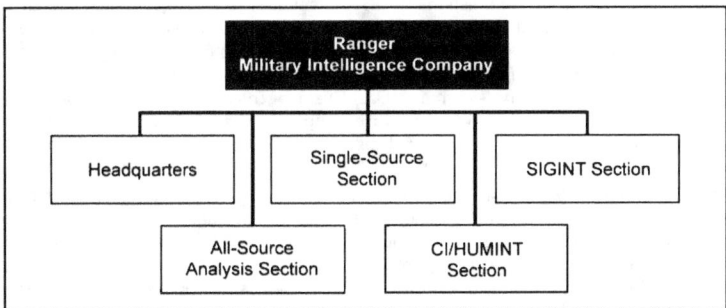

Figure 6-2. Ranger Military Intelligence Company organization

Collection Management and Dissemination Section

6-9. The collection management and dissemination section is part of the HQ element. It is comprised of an officer, a noncommissioned officer, and two DA civilians who function as Webmasters. The section synchronizes and monitors the intelligence staff's collection requirements with internal and external collection sources. The collection manager plans and coordinates strategic surveillance and reconnaissance requirements. Duties include conducting research and exploiting national databases to facilitate an all source analysis approach and provide the most current data for target intelligence packages. The Webmasters are responsible for designing and maintaining the regiment's JWICS intelligence page to disseminate intelligence products to the Ranger force.

All-Source Analysis Section

6-10. The all-source analysis section (Figure 6-3, page 6-4) uses systems and software to graphically display the enemy situation and produce target folders to support operations. The section consists of three order of battle teams and the Home Station Operations Center intelligence support team. They also maintain databases of intelligence information, conduct pattern analysis, exploit national databases for needed intelligence, and search for relevant message traffic to build and display the enemy picture. This section produces intelligence products tailored for dissemination down to the company level to facilitate operator tactical usability.

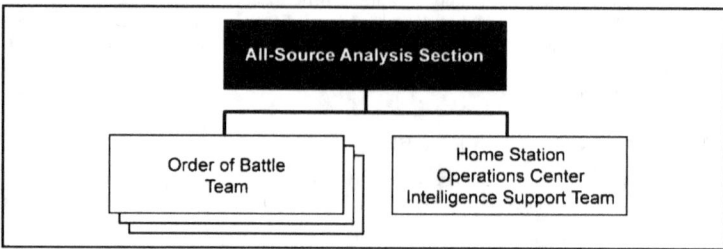

Figure 6-3. All-source analysis section

Single-Source Section

6-11. The single-source section (Figure 6-4, page 6-5) is comprised of three subordinate sections responsible for individual intelligence disciplines:

- *Imagery analysis section.* The imagery analysis section is responsible for the acquisition, exploitation, and dissemination of imagery, motion video, and geospatial data and products from national and tactical assets in direct support of missions. Imagery analysts are also responsible for battle damage assessment analysis using current imagery. The imagery analysts use the EIW to perform their imagery analysis mission. The imagery analysis section is augmented with a civilian imagery analyst from the NGA. This imagery analyst provides a direct link to the NGA supporting imagery support staff, as well as mentorship to the imagery analysts and instruction on use of imagery software and analysis. The NGA imagery analyst is capable of deploying with the unit.
- *Terrain analysis section.* The terrain analysis section performs cartographic and terrain analysis. This section provides the ground maneuver units with detailed analysis of terrain, to include lines of communication, terrain elevation, and natural or man-made obstacles. They predict terrain and weather effects as applied to mission command, communication, and computer and intelligence systems. The terrain analysis section is equipped with the Digital Topographic Support System-Deployable. The terrain analysis section is augmented with a civilian geospatial analyst from the NGA. The geospatial analyst provides mentorship to the imagery and terrain analysts, instructs them on the use of geospatial software, and generates specific GEOINT products to support the unit. The geospatial analyst is capable of deploying with the unit.
- *Technical Control and Analysis Element (TCAE).* The TCAE performs technical analysis. SIGINT analysts in the TCAE operate SIGINT programs accessed though the JWICS and the NSA network. The TCAE analyst uses a combination of Army-common and special operations intelligence systems. These systems are designed to be interoperable with theater intelligence systems and national assets, such as the Tactical Exploitation of National Capabilities program.

Counterintelligence/Human Intelligence Section

6-12. The CI/HUMINT section includes a CI team and two tactical HUMINT teams that are commanded and controlled by the operations management team. The operations management team is responsible for the overall management of all CI and HUMINT teams operating in the regiment's area of responsibility. CI teams are responsible for spotting, accessing, and recruiting human sources to collect regimental priority intelligence requirements. The tactical HUMINT teams have the capability to operate a temporary detainee holding area and conduct interrogations of prisoners of war.

Intelligence Support to Rangers

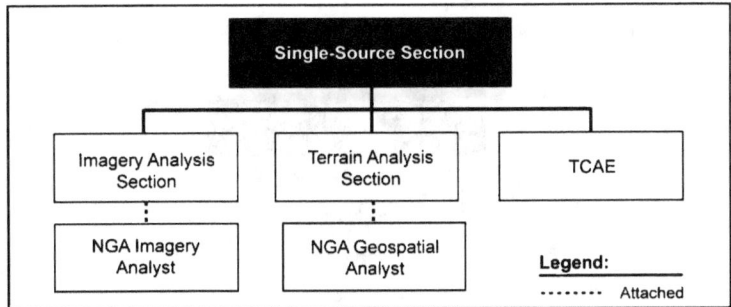

Figure 6-4. Single-source section

Signals Intelligence Section

6-13. The SIGINT section consists of three operational SIGINT teams that conduct ground-based SIGINT collection to find and fix the location of the enemy, and to provide indications and early warning of enemy intent or actions. The SIGINT analysis section gathers, sorts, and analyzes intercepted messages from the operations SIGINT teams to isolate valid message traffic. Additionally, it evaluates intelligence data from SIGINT reports released from national agencies. The information gathered is used to identify actionable targets.

RANGER RECONNAISSANCE COMPANY

6-14. The Ranger reconnaissance company (Figure 6-5, page 6-6) consists of a HQ element, which includes a selection and training team and an operations section, and six Ranger reconnaissance teams. The primary mission is to conduct all forms of reconnaissance and surveillance, and limited direct action to support Ranger Regiment missions. The Ranger reconnaissance company is under the control of the regimental commander and regiment's operations officer for planning its missions. Ranger reconnaissance teams can operate directly for a Ranger battalion commander or in support of the regiment or higher HQ. The Ranger reconnaissance teams give the regiment the capability to conduct operational preparation of the environment and answer intelligence requirements. The Ranger reconnaissance company—

- Fulfills target area surveillance missions in an area before committing other Ranger or USASOC elements to the operation.
- Engages hostile targets with direct fire, indirect fire, and demolitions.
- Conducts limited terminal guidance operations.
- Conducts prestrike and poststrike surveillance on critical nodes for battle damage assessment requirements.
- Is specially trained to provide information on threat order of battle and target sites, and to conduct route and limited chemical, biological, radiological, nuclear, and high-yield explosives reconnaissance.
- Conducts pathfinder operations to reconnoiter, select, clear, and prepare landing and drop zones.
- Can conduct autonomous tactical operations for up to five days in denied areas.

6-15. The Ranger reconnaissance company conducts infiltration and exfiltration using several methods. These methods include the following:

- Military free fall, to include high-altitude low-opening and high-altitude high-opening parachute techniques.
- Vehicular (standard/nonstandard).
- Low-visibility actions.

Chapter 6

- Small boat.
- Scout swimmer.

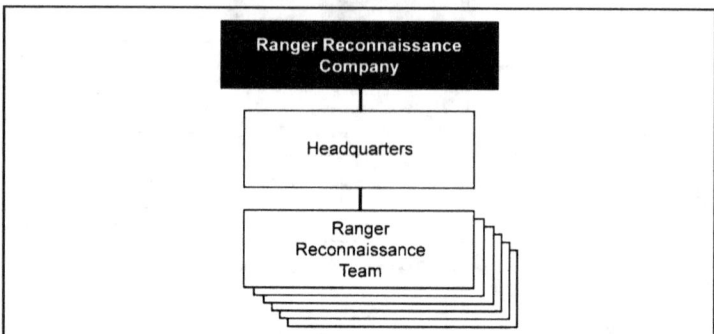

Figure 6-5. Ranger reconnaissance company organization

BATTALION INTELLIGENCE STAFF SECTION

6-16. The Ranger battalion intelligence staff section (Figure 6-6, page 6-7) consists of three officers—the intelligence officer and two assistant intelligence officers—and five enlisted Soldiers. This section has the capability to collect and analyze information. Its mission is to support the battalion commander with basic intelligence, database maintenance, collection management, analysis, and tactical intelligence production and dissemination for battalion operations. The Ranger Military Intelligence Company or external assets, as required, augment the battalion intelligence staff section. Additionally, the Ranger battalion intelligence staff—

- Tasks, through the operations staff battalion, elements to perform combat intelligence missions supporting battalion operations.
- Conducts intelligence training for battalion elements.
- Supports the planning, coordination, and execution of Ranger target rehearsals.
- Briefs and debriefs reconnaissance teams.
- Identifies, confirms, and coordinates priorities for input to regimental geographic area requirements for GI&S products.
- Is responsible for personnel and physical security.

BATTALION RECONNAISSANCE PLATOON

6-17. Each Ranger battalion has a battalion reconnaissance platoon. A battalion reconnaissance platoon consists of a HQ element and four Ranger reconnaissance sections with six personnel each. Its primary mission is to conduct tactical reconnaissance and surveillance and limited direct action in support of the battalion. The battalion reconnaissance platoon is under the control of the battalion commander and operations officer for planning its missions. The battalion reconnaissance platoon provides the battalion commander with necessary organic reconnaissance and surveillance to answer his priority intelligence requirements.

OTHER ORGANIC SUPPORT

6-18. In addition to the support provided by the Ranger Regiment, medical and fire support personnel provide other organic support. The regimental and battalion surgeons are a source of medical and chemical, biological, radiological, nuclear, and high-yield explosive intelligence concerning possible deployment locations. They also provide valuable information on disease and health conditions in the area of operations. Fire support personnel are able to serve as a conduit for information collected by units providing supporting fires. This information may include detections by the counter battery radar of an artillery unit or observations by pilots flying close air support missions.

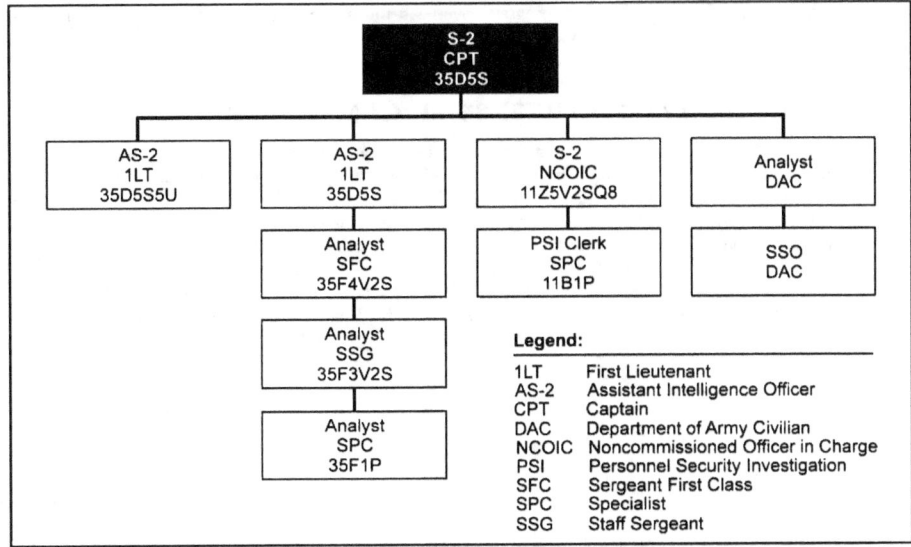

Figure 6-6. Ranger battalion S-2 organization

NONORGANIC INTELLIGENCE SUPPORT

6-19. Intelligence support for Ranger operations is specialized and sensitive. Nonorganic intelligence support to the Ranger Regiment is discussed in the following paragraphs.

SUPPORT FROM HIGHER HEADQUARTERS

6-20. The USASOC intelligence staff's Intelligence Operations Division coordinates with USSOCOM Center for Special Operations, Special Operations Command Joint Intelligence Center, and theater and national intelligence agencies for the information and intelligence needs of the Ranger Regiment.

6-21. When deployed, supported units provide intelligence support to the Ranger Regiment. The supported unit's intelligence staff provides target-specific intelligence in the form of a target intelligence package. The supported unit's intelligence staff provides multidiscipline CI and electronic warfare support to the Ranger force. The Ranger Regiment normally places an intelligence liaison at the supported unit or the appropriate intelligence production facility to—

- Ensure the intelligence needs of the Ranger force are relayed to the appropriate intelligence processing center.
- Ensure the resulting analysis is based on the specific needs of the Ranger force commander.
- Coordinate with the corps or echelons above corps targeting center when the Ranger force is under operational control of a conventional force.
- Coordinate between the regiment, the corps or echelons above corps intelligence node, and the targeting center to effectively employ Ranger forces.

Chapter 6

RECONNAISSANCE AND SURVEILLANCE SUPPORT

6-22. When deployed, additional reconnaissance and surveillance assets are provided by the supported unit to assist the Ranger Military Intelligence Company in detection and target refinements. Although the Ranger Regiment has a tactical UAS program, overhead full-motion video by national assets, such as the Predator UAS, are normally provided by the supported unit. Additional collection assets normally requested through the supported unit are national SIGINT collection and national imagery collection.

RANGER SUPPORT TO THE INTELLIGENCE PROCESS

6-23. The robust capability of the Ranger reconnaissance company to conduct reconnaissance through its six reconnaissance teams not only provides the regiment with organic reconnaissance and surveillance capabilities but complements the reconnaissance and surveillance capability of the force in theater. Rangers are often the first to encounter the enemy and can confirm or deny friendly assessments of adversary organization, equipment, capabilities, and morale. They can bring back captured adversary equipment for evaluation and report on the effectiveness of friendly weapons on threat systems. Rangers can also provide real-time assessments of the target area civilian population's morale and their physical disposition for use in MISO and CAO plans. Ranger intelligence staffs must be proactive in debriefing Rangers to ensure this valuable information enters the intelligence process.

Chapter 7
Intelligence Support to Special Forces

This chapter describes SF activities across the range of military operations. It focuses on SF principal tasks, intelligence requirements, typical organization, intelligence support, and SF unit input to the intelligence process. The chapter highlights the challenges in providing intelligence support to SF operational detachments. The fundamental purpose of SF intelligence operations is to support commanders of SF units with timely, relevant, accurate, and predictive information on the enemy and the environment. This support enables commanders to conduct successful worldwide operations, whether deployed as individual teams or as part of a larger element in support of GCCs, U.S. Ambassadors, and other government agencies.

MISSION

7-1. SF units conduct a variety of special operations by performing nine principal tasks. The tactical actions associated with performing these tasks may often have operational or strategic effects. SF units conduct operations that require flexible and versatile forces that can function effectively in diverse and contradictory environments. SF Soldiers apply this flexibility and versatility through their nine principal tasks: unconventional warfare, foreign internal defense, counterinsurgency, security force assistance, direct action, special reconnaissance, counterterrorism, counterproliferation of weapons of mass destruction, and information operations.

INTELLIGENCE REQUIREMENTS

7-2. The intelligence required to support SF units differs somewhat from the composition of other operating forces. The specificity and level of detail required to support all SF principal tasks is typically greater than that required by conventional units. Due to the operational or strategic significance of the missions executed by SF units, higher levels of intelligence support are frequently required to plan, execute, and assess SF missions. The requirement for national-level intelligence support is the norm with these missions.

INTELLIGENCE ORGANIZATION

7-3. Operations conducted by SF units require leveraging organic intelligence support with analysis, collection management, security, weather, terrain, and communications capabilities tailored to support their supported geographic combatant command requirements across the range of military operations. SF unit intelligence assets are organized according to operational and analytical needs, and are located at the group, battalion, company, and operational detachment levels. Unit intelligence staffs and military intelligence detachments (MIDs) are key components of the organic intelligence capability at the group and battalion level. The staff plans, organizes, directs, coordinates, and controls collection operations, while the MID commander executes the directives.

STANDING INTELLIGENCE ORGANIZATIONS

7-4. In garrison, SF groups organize military intelligence personnel into a distinctly different organization from that used during deployment. The following sections detail standing and deployed intelligence organizations.

Chapter 7

Group Intelligence Section

7-5. The group intelligence section consists of an intelligence officer, an assistant intelligence officer, an SF warrant officer, and a noncommissioned officer in charge. Figure 7-1 depicts the group intelligence section.

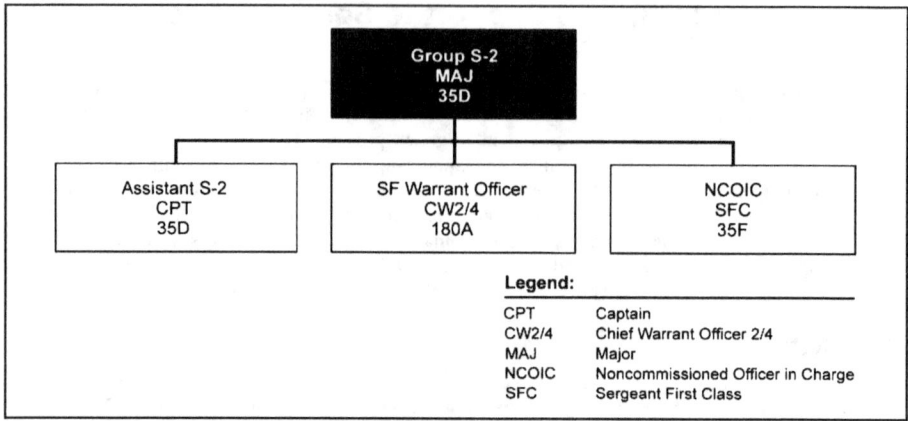

Figure 7-1. Special Forces group S-2 section

Intelligence Officer

7-6. The intelligence officer is a military intelligence major who is the primary staff officer for all aspects of intelligence, CI, and security support in garrison and while deployed. The intelligence officer plans, coordinates, and directs all group-level intelligence collection, analysis, and production, and approves all products before dissemination. The intelligence officer develops and recommends priority intelligence requirements and information requirements for approval by the group commander and maintains the group collection plan with the assistance of the group MID collection management and dissemination team. The intelligence officer directs all group-level intelligence collection operations involving organic or attached assets. The intelligence officer identifies, confirms, and coordinates priorities for input to regimental geographic area requirements for GI&S products.

7-7. The intelligence officer identifies the need for collateral and compartmentalized intelligence communications support and intelligence automated data processing systems support. The intelligence officer works with the signal staff officer in planning and coordinating systems support. The intelligence officer coordinates GI&S requests and products, and then disseminates GEOINT. The terrain team produces GEOINT products from the fusion of the geospatial and weather information. The intelligence officer conducts or coordinates for a wide variety of CI activities in support of the protection warfighting function. When the group forms a SOTF, the intelligence officer serves in the operations center under the staff supervision of the operations center director. The intelligence officer is responsible for the group's information security, information systems security, personnel security, and special security programs. He establishes and ensures that the group maintains an intelligence training program that trains military intelligence and SF personnel. The intelligence officer coordinates with the group's MID commander to establish a tactical SCIF when required. He coordinates tasking and operational control of MID assets with the requesting SF unit and the MID commander.

Assistant Intelligence Officer

7-8. The assistant intelligence officer is a military intelligence captain who is the focal point for collection operations and current intelligence at group level. When a SOTF is established, the intelligence officer determines the assignment of the assistant intelligence officer. The assistant intelligence officer potentially serves as chief of the current intelligence branch (collocated with the operations section in the operations center), the JISE director, or the collection management and dissemination chief located in the SCIF. Responsibilities of the assistant intelligence officer include maintaining the current intelligence estimate and other current situation intelligence products, as well as coordinating with the intelligence warrant officer to monitor the status of all deployed intelligence collection assets. The assistant intelligence officer also develops taskings for the staff weather officer and the USAF SOWT. The assistant intelligence officer serves as the acting intelligence officer or senior intelligence officer in the absence of the intelligence officer.

Intelligence Warrant Officer

7-9. The group intelligence warrant officer is an SF warrant officer who is the assistant intelligence officer for plans and targeting support at group level. When a SOTF is established, the group intelligence warrant officer serves as chief of the plans and targeting support branch, which collocates with the operations plans branch of the operations staff. He works with the special plans officer in the operations staff and the group CI technician to coordinate HUMINT collection and integration. Additionally, he is responsible for coordinating all aspects of intelligence support for future plans and targeting. For this reason, he must be completely familiar with all of the intelligence functions of the group MID. As the senior SF Soldier of the intelligence staff, the group intelligence warrant officer works with and helps the group intelligence officer and MID commander develop, focus, and coordinate emerging intelligence-related subjects and training for personnel holding military occupational specialty 18F or 180A.

Noncommissioned Officer in Charge

7-10. The group intelligence staff noncommissioned officer in charge is an all-source intelligence analyst who is in charge of the noncommissioned officers and enlisted personnel within the group intelligence staff. When a SOTF is established, he ensures that the intelligence staff is integrated with the other staff sections. In addition, he aids the intelligence officer in IPB planning, collection management, and targeting. The noncommissioned officer in charge serves as the chief advisor to the intelligence officer on all matters related to the group's physical, document, information, and automation systems security programs. The noncommissioned officer in charge supervises readiness training funds management and training, the special security office, and group-wide 18F intelligence training.

Group Military Intelligence Detachment

7-11. The MID contains most of the group's single-source and all-source analysis capability. The detachment is responsible for—

- Collection management.
- Single-source collection.
- All-source integration of single-source information.
- Analysis, production, and dissemination of finished intelligence products.
- Control and management of the SCI local area network team.

7-12. The MID (Figure 7-2, page 7-4) consists of four subordinate sections. Each section has different functional responsibilities.

Chapter 7

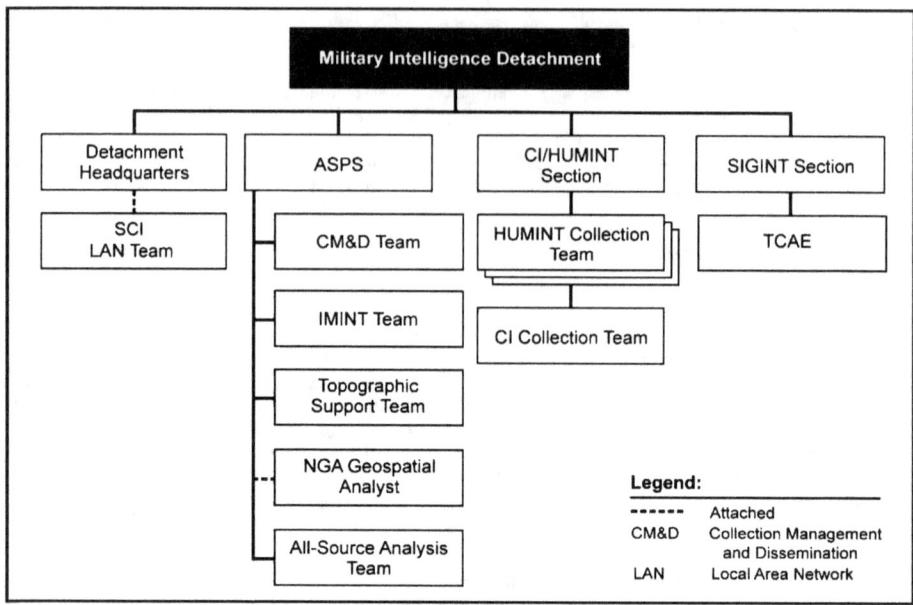

Figure 7-2. Special Forces group military intelligence detachment

Detachment Headquarters

7-13. The MID commander directs the efforts of the analytic elements of the detachment and mission support for the intelligence collection activities of the CI and HUMINT teams deployed forward. He works with the group intelligence officer in garrison, and directly for the group intelligence officer when task-organized for deployment, helping to develop the group's concept for intelligence operations. The commander implements the MID task organization, deploying collection assets and analytic augmentation to subordinate units as directed by the group commander. The detachment commander is responsible for individual and collective training for military intelligence specialties and for property accountability of equipment unique to the MID. However, the MID does not have the organic maintenance, supply, or administrative personnel to operate independently. The MID must rely on and coordinate closely with the group support company for administrative and sustainment support.

Sensitive Compartmentalized Information Local Area Network Team

7-14. The SCI local area network team terminates SOT-A communications and operates and maintains SCI communications and automated data processing systems linking a SOTF into the intelligence system. The SCI local area network team maintains SCI communications with higher and adjacent HQ and a SOTF. It assists the MID in establishing and maintaining SCI communications links using the TS-LITE.

All-Source Production Section

7-15. The All-Source Production Section (ASPS) is the hub of the MID. It is directed by an all-source intelligence warrant officer and responds to all-source intelligence support taskings from the group intelligence officer. The ASPS provides in-depth all-source analysis, production, and collection management support to the group and near-real-time indications and warnings overwatch for deployed assets. All members of the ASPS contribute to the preparation of the intelligence estimate and its corresponding annexes. The ASPS produces and enhances target intelligence packages for subordinate units.

7-16. The ASPS has four subordinate sections with specific functions—the collection management and dissemination team, the IMINT team, the topographic support team, and the all-source analysis team. The ASPS may be further subdivided into analysis teams focused on specific areas of operations.

Collection Management and Dissemination Team

7-17. The collection management and dissemination team of the ASPS receives, validates, and prioritizes all requests for information from subordinate units and assigns them to the appropriate section. The tasked section informs the ASPS chief upon priority intelligence requirement satisfaction and recommends new priority intelligence requirements, information requirements, and specific information requirements for submission to the intelligence staff. Based on guidance from the group intelligence officer, the collection manager reports intelligence to higher or lateral HQ, and coordinates dissemination of intelligence to subordinate and attached units.

7-18. A noncommissioned officer normally leads the collection management and dissemination team; however, it could be task-organized by the group intelligence officer to have an available officer fill the position as it is a very demanding job requiring interfacing with multiple HQ in the theater. The team may be augmented with other intelligence personnel from within the MID. The team develops and helps implement a collection plan that supports mission analysis and planning, targeting, indications and warnings for deployed elements, and security requirements. Personnel from the single-source intelligence disciplines subordinate to the MID help the collection management and dissemination team with these tasks, as required. Collection management and dissemination personnel work with the group intelligence staff to develop the group's collection plans. As part of this effort, the team develops information objectives and collection emphasis. The team also works with the SIGINT and CI/HUMINT sections and the IMINT team to develop SIGINT amplifications and end products review, source-directed requirements, imagery reconnaissance objective lists, and taskers. The collection management and dissemination team works with the group's special security officer to maintain the collateral recurring document listing, SCI recurring document listing, and the statement of intelligence interest. The collection management and dissemination team often conducts UAS mission management in conjunction with the IMINT team.

Imagery Intelligence Team

7-19. The IMINT team of the ASPS maintains comprehensive historical and current mission area imagery files in hard copy and digitized form. The imagery analysts of the team use their EIWs, imagery product library system, and imagery products from higher HQ to provide in-depth imagery analysis and production support to the group. This support includes detailed terrain and facility analysis, annotated prints and target graphics, current Secondary Imagery Dissemination System imagery of targets and key terrain, battle damage assessment, and highly accurate imagery-based object measurement support.

Topographic Support Team

7-20. The topographic support team provides timely, accurate knowledge of the area of operations and terrain visualization to the commander, his staff, and subordinate units throughout the range of special operations. The team collects and processes military geographic information from remotely sensed imagery, digital data, intelligence data, existing geospatial products, and other collateral data, to include geospatial data from the field. The team can print or duplicate terrain analyses and nonstandard map products for the unit.

National Geospatial-Intelligence Agency Geospatial Analyst

7-21. The NGA geospatial analyst supports the SF group by providing a direct link to NGA products and support. The geospatial analyst also provides mentorship to the imagery and terrain analysts, instructs them on the use of geospatial software, and generates specific GEOINT products to support the unit mission. The geospatial analyst is capable of deploying with the unit.

Chapter 7

All-Source Analysis Team

7-22. All-source analysts within the all-source production center analyze the area of operations, conduct IPB, nominate targets, and conduct battle damage assessment. They prepare and continually update the current intelligence situation map. They maintain and use the unit intelligence reference files and databases, including the SOCRATES. They help the collection management and dissemination team develop the collection plan and identify new requirements to close any intelligence gaps. They prepare intelligence summaries and responses to requests for information and provide intelligence briefings, as required.

Counterintelligence/Human Intelligence Section

7-23. The CI/HUMINT section provides a large number of tactical CI and HUMINT collection, analysis, production, and operational support activities. A CI technician leads the CI/HUMINT section. The section consists of CI agents trained in CI functions and HUMINT collectors operating in two-man teams. They are normally task-organized by language capability for the mission. CI and HUMINT personnel use the CHIMS/CHATS and the Counterintelligence and Operations Workstation to perform their tasks. They are issued the CHIMS.

7-24. The CI/HUMINT section produces intelligence information reports and other formatted reports to disseminate the results of CI and HUMINT collection and liaison activities. The section uses locally collected intelligence and theater CI products to assess a wide variety of threats for deployed units in the communications zone or a rear area. These threats include, but are not limited to, foreign intelligence services, insurgents, terrorists, criminals, dissident political factions, and anti-U.S. elements of the civilian population. The CI/HUMINT team also prepares and maintains the unit's CI estimate, as well as comprehensive base defense, communications zone, and rear area IPB products. These products permit the timely provision of indications and warnings information and predictive intelligence on significant security threats. CI and HUMINT analysts develop friendly force profiles and identify friendly vulnerabilities.

7-25. In response to the situation and collection taskings from the collection and management team and the MID commander, the CI/HUMINT section plans, coordinates, and conducts CI and local law enforcement liaison, EPW interrogation, document exploitation, overt collection, and refugee and evacuee debriefings. With approval, tactical HUMINT teams conduct CI force protection source operations. Teams process plans, coordinate financial and administration support, and ensure regulatory intelligence oversight compliance. The teams recommend essential elements of friendly information and threat countermeasures. They continually assess the effectiveness of the base OPSEC countermeasures and base security plans by establishing and updating threat vulnerability assessments. The CI and HUMINT teams support the OPSEC plan by providing TARP briefings and limited investigations. The teams also provide CI and security site surveys in support of unit relocations within the country and theater, as required, and conduct debriefings of friendly units.

7-26. The CI/HUMINT section conducts liaison activities with HN government (national and local) law enforcement, intelligence, and security officials. It also establishes, maintains, coordinates, and deconflicts CI and HUMINT operations and activities with the higher and adjacent intelligence staffs, as required. All CI and HUMINT operations require coordination and deconfliction through the S-2X.

Signals Intelligence Section

7-27. The SIGINT section conducts single-source collection, collection management, and analysis. The section is led by a SIGINT warrant officer and consists of the TCAE. The TCAE publishes reports for the SOT-A and nonorganic SIGINT systems into the national SIGINT system. The TCAE performs all SIGINT operational tasking authority functions on behalf of the group commander.

Technical Control and Analysis Element

7-28. The TCAE performs technical analysis. The TCAE analyst uses a combination of Army and special operations intelligence systems. These systems are designed to be interoperable with theater intelligence systems and national assets, such as the Tactical Exploitation of National Capabilities program. The

Intelligence Support to Special Forces

Tactical Exploitation of National Capabilities program either maintains or has access to the SIGINT databases for selected countries in the area of operation. SIGINT analysts in the TCAE operate SIGINT-related programs accessed though the JWICS and the NSA network. The TCAE provides critical combat and technical information to the all-source analysts within the ASPS to support situation development, intelligence summaries, and intelligence briefings to the commander and staff. Additionally, the TCAE—

- Determines specific SIGINT collection taskings from stated intelligence requirements received by the collection management and dissemination team.
- Monitors the effectiveness of collection efforts, redirecting coverage and providing technical feedback and support to the SOT-A and SIGINT personnel attached to SF battalion units.
- Establishes, updates, and maintains the group's intercept tasking database.
- Ensures adherence to all legal aspects of SIGINT operations.
- Issues all appropriate SIGINT operational tasking authority reports.

DEPLOYED S-2 ORGANIZATIONS

7-29. The following paragraphs represent an abbreviated list of the functions of the deployed intelligence staff section under the direction of the JISE director and the SOTF SCIF. Under the leadership of the intelligence officer, the intelligence section is the focal point for all-source intelligence production, collection management, and synchronization of the intelligence process. These are examples and should be interpreted as such. Each group intelligence officer may task-organize his SCIF as required by mission, enemy, terrain and weather, troops and support available, time available, and civil considerations.

Intelligence Section

7-30. Under the staff supervision of the unit intelligence staff, the operations center's intelligence section is the focal point for all-source intelligence production and collection management. This section normally consists of intelligence personnel, either from the group staff or the group MID. Members from the SOWT work in the operations center under the direction of the group operations officer (Figure 7-3).

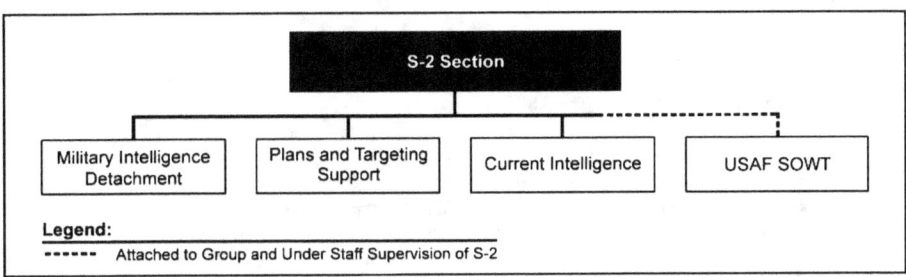

Figure 7-3. Special operations task force operations center S-2 section

Military Intelligence Detachment

7-31. Under the supervision of the group intelligence officer, the MID is the focal point for all-source intelligence production, collection management, SIGINT, HUMINT, and GEOINT.

Plans and Targeting Support Branch

7-32. The consolidated plans section maintains operation plans for the entire operations center. The intelligence staff representative to the consolidated plans section coordinates and plans intelligence support for future and contingency operations. The intelligence staff representative locates with the operations or plans branch or the consolidated plans section.

Chapter 7

Current Intelligence Branch

7-33. The intelligence staff current intelligence branch coordinates the daily operations of the intelligence staff and MID. The current intelligence branch physically works in the operations center in conjunction with the operations staff. The branch is responsible for providing intelligence expertise to the operations officer for all immediate and ongoing operations.

USAF Special Operations Weather Team

7-34. The USAF SOWT is attached to SOTFs and is under staff supervision of the task forces' intelligence staff officer. The USAF staff weather officer serves as a special staff officer to the task force commander on all weather matters. The team should locate within the operations center to facilitate the exchange of information to other intelligence sections and the operational sections. The detachment provides—

- Current and forecast weather and light data.
- Information and analysis to support the determination of weather effects upon operations.
- Climatic analysis studies supporting all group missions.
- Forward area limited-observing program and remote weather sensor training and mission support information down to the SF operational detachment level.

7-35. When approved by the commander, the SOWT may attach team members to subordinate elements and units to gather critical weather observations from denied territories and regions for which there are underdeveloped databases.

Military Intelligence Detachment

7-36. The basic organization of the MID in a deployed status differs little from when at home station (Figure 7-4, page 7-9). The principal difference is the command relationship. At home station, the MID has no operational command relationship with the intelligence staff, whereas for a deployment, the detachment falls under the operational control of the intelligence staff officer to maintain continuous administrative functions.

All-Source Production Section

7-37. The ASPS performs IPB, develops target data, and consolidates information from all sources to meet the commander's mission needs. The ASPS—

- Processes, correlates, and integrates all-source intelligence in response to taskings from the collection management and dissemination section.
- Is the focal point for all situation and target development.
- Develops and maintains the unit's intelligence database, including the intelligence journal, order of battle information, IPB products, targeting data, and the situation map.
- Monitors the collection plan and recommends revisions to close identified gaps.
- Receives and processes intelligence products and combat information from higher, lower, and adjacent commands.
- Prepares intelligence estimates, reports, summaries, and briefs, as required.

Collection Management and Dissemination Team

7-38. The collection management and dissemination team of the ASPS formulates detailed collection requirements and tasks or requests the collection of required information. Normally collocated with the ASPS, the collection management and dissemination team—

- Performs intelligence collection management for the intelligence operations branch.
- Obtains the command's approved requirements from the intelligence staff officer, prioritizes them based on command guidance, and translates them into the collection requirements.
- Prepares and continuously updates the unit collection plan and forwards the unit's information requirements to the higher command.

- Tasks organic and attached collection assets.
- Requests, through the operations staff, intelligence collection mission tasking of organic teams or other subordinate assets.
- Disseminates combat information and intelligence within the command and to higher, lower, and adjacent units.
- Conducts reconnaissance and surveillance mission management, to include battle damage assessment.

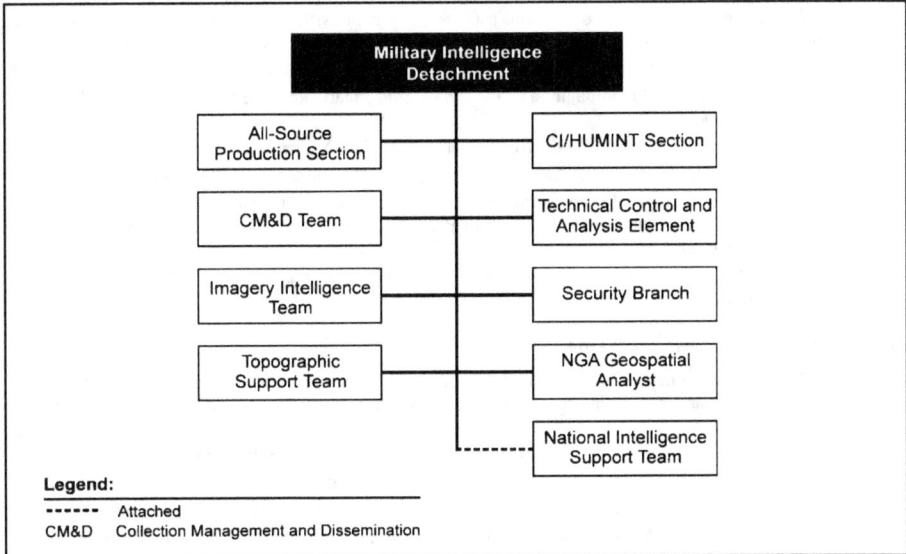

Figure 7-4. Special operations task force operations center military intelligence detachment

Imagery Intelligence Team

7-39. Imagery analysts extract information and develop intelligence mainly from imagery collected from aerial and Tactical Exploitation of National Capabilities program sensors. Within SF units, imagery analysts are assigned to the group and battalion MIDs. A thorough knowledge of threat tactics and ground order of battle enables them to recognize, identify, locate, describe, and report information on objects, activities, and terrain from a variety of imagery products. Imagery analysts make associations between visible objects and configurations and analyze the results to determine strength, disposition, and enemy capabilities. The IMINT team—

- Analyzes imagery and reports specific information on threat operations, activities, dispositions, sustainment, communications, installations, and civilian activities, and their possible effect on operations.
- Prepares and maintains imagery prints to supplement and update maps for operational planning.
- Extracts information from imagery in direct support of unit mission area analysis.
- Supports combat assessment.
- Prepares (along with the terrain team) mosaics and target terrain models to support operational planning.
- Accomplishes imagery exploitation, reporting, and production for SOTF requirements from joint or theater imagery exploitation activities.

- Prepares and maintains an imagery database for the unit's specified mission areas.
- Provides technical assistance to the unit collection management and dissemination section, as required.

Topographic Support Team

7-40. The topographic support team consists of two topographic analysts. Their primary equipment is the DTSS-D, which provides terrain analysts with combat terrain information to support terrain visualization, IPB, and mission command. Topographic analysts use the DTSS-D to generate tactical decision aids, process digital satellite imagery from various sources, provide full-size map reproduction, and use Web-based technology to share terrain information and data. The topographic support team—

- Analyzes collected military geographic information to produce tactical decision aids.
- Performs digital manipulation of topographic information by querying, viewing, evaluating, and downloading digital data.
- Predicts terrain and weather effects as applied to communications systems support.
- Performs database management for the storage of aerial photographs, maps, and digital databases.
- Performs cartographic and terrain analysis duties and collects and processes military geographic information from remotely sensed imagery, digital data, intelligence data, existing topographic products, and other collateral data sources.
- Draws, digitizes, and scans cultural, topographic, hydrographic, and other features on overlay or into digital formats.

Counterintelligence/Human Intelligence Section

7-41. The primary function of the CI/HUMINT section is to conduct CI and HUMINT collection operations. Additional capabilities include analysis, force protection, OPSEC, and deception programs. The CI/HUMINT section—

- Supports the unit's personnel security and information security functions.
- Conducts liaison with other U.S. and security organizations located near the forward bases.
- Conducts CI analysis to support ASPS security and situation and target development efforts.
- Develops detailed assessments of foreign intelligence and security threats near SOTFs and in SF areas of operations.
- Recommends appropriate measures to reduce friendly vulnerabilities.
- Evaluates, if possible, the effectiveness of measures implemented to correct identified friendly vulnerabilities.
- Briefs deploying teams on the latest threat data.
- Provides teams with technical advice and assistance to prepare them to establish and operate during long-term missions.
- Supports operations by determining foreign vulnerabilities to deception.
- Provides the operations staff officer with recommendations for deception measures and evaluates their effectiveness through CI analysis.
- Requests external support, when necessary, to evaluate foreign reactions to friendly deception operations.
- Uses current intelligence automated data programs to exchange information with existing higher, adjacent, and subordinate element databases.

7-42. The HUMINT analysis team is the focal point for all HUMINT reporting and operational analysis. Other functions include the following:

- Determines gaps in reporting and coordinates with the collection manager to cross-cue other intelligence assets.
- Produces and disseminates HUMINT products.
- Uses analytical tools to develop long-term analysis and provides reporting feedback that supports the HUMINT teams.

7-43. The group MID has three two-man HUMINT teams that may be task-organized, as required. The teams may be attached to the subordinate battalions or a joint interrogation facility. HUMINT team activities include—
- Interrogating EPWs and detainees.
- Debriefing detainees, returned U.S. personnel, and other persons of intelligence interest to the supported commander.
- Exploiting documents that appear to satisfy the supported commander's priority intelligence requirements directly.
- Conducting (when directed) CI force protection source operations, as authorized in the theater CI plans.

7-44. The primary system the CI/HUMINT section uses in the execution of its duties is the CHIMS/CHATS, an information management system for use in the field. The CHIMS/CHATS provides the team leader the capability to collect, process, and disseminate information obtained through HUMINT operations. The CHIMS/CHATS consists of two suitcase-sized containers that hold the following items:
- Laptop computer with removable hard drive.
- Compact disc and digital video disc read-only memory drives.
- Fingerprint verifier.
- Secure telephone unit.
- Single-channel ground and airborne radio system and A/N PSC-5 series radio interface cables.
- Support kit with universal power converters.
- Color printer.
- Scanner.
- External disk drives.
- Digital camera.
- Modem.
- Global positioning system receiver.

Note: In addition to the typical commercial Windows-based software assortment, the laptop computer contains software programs that assist in link analysis, association matrices, link diagrams, and photo editing.

7-45. Another tool available to the CI/HUMINT section is the Individual Tactical Reporting Tool. The Individual Tactical Reporting Tool provides CI and HUMINT teams the capability to collect, process, and disseminate tactical intelligence obtained through investigations, interrogations, collection operations, and document exploitation. The Individual Tactical Reporting Tool is an entry-level device for reporting to the CHIMS/CHATS, through commercial- and government-developed software designed to receive, process, and store formatted messages, digital maps, and imagery.

Technical Control and Analysis Element

7-46. The group TCAE performs SIGINT and electronic warfare management functions, and provides control of SOT-As. Specific functions of the TCAE include—
- Producing the SIGINT collection plan.
- Providing centralized technical control and collection tasking authority over deployed SOT-As.
- Analyzing and correlating intercepted SIGINT traffic from the SOT-As with data from other sources and then passing these products to the ASPS and higher-echelon TCAEs or SIGINT processing centers.
- Developing and maintaining the SIGINT technical database and the electronic order of battle database.

Chapter 7

- Interfacing with theater and national intelligence systems to complete the integration of technical data generated by tactical units with the technical data produced by the NSA and pulling technical data required for SIGINT and electronic warfare operations.
- Providing technical support (such as SIGINT technical data) to SOT-As, as required.
- Performing all functions of the SIGINT collection management authority. The TCAE officer in charge performs these functions.
- Publishing reports from SOT-As and nonorganic SIGINT systems into the national SIGINT system.

Note: CONUS-based TCAEs interface directly with the Army TCAE during peacetime for all technical support requirements and readiness reporting.

Security Branch

7-47. The S-2 security branch develops unit personnel, information, automated data processing, and physical security programs, and supervises their implementation. It may also be responsible for physical security, badge and pass procedures, and courier orders, and serve as the focal point for security clearances. Under its staff supervision, the CI/HUMINT section of the MID provides CI support to the operations staff's OPSEC program and to deception planning. The security branch of the intelligence staff establishes and administers security procedures for the tactical SCIF. It establishes emergency procedures for the removal, protection, and destruction of classified materials.

National Geospatial-Intelligence Agency Geospatial Analyst

7-48. The NGA geospatial analyst supports the SF group by providing a direct link to NGA products and support. The geospatial analyst also provides mentorship to the imagery and terrain analysts, instructs them on the use of geospatial software, and generates specific GEOINT products to support the unit mission. The geospatial analyst is capable of deploying with the unit.

National Intelligence Support Team

7-49. A JSOTF intelligence staff may request the deployment of a NIST. A NIST is comprised of intelligence experts from DIA, CIA, NGA, NSA, and other agencies as required to support the specific needs of the mission.

Other Intelligence Assets

7-50. The group and battalion medical sections are excellent sources of information about the operational area, its associated health threats, and the medical personnel and facilities for these locations. The medical area studies developed by Army Medical Department intelligence personnel are an additional source of information.

Battalion Intelligence Staff

7-51. The battalion intelligence staff officer (Figure 7-5, page 7-13) is the primary staff officer for all aspects of intelligence, CI, and security support in garrison and while deployed. The intelligence staff officer plans, coordinates, and directs all battalion-level intelligence collection, analysis, and production. The intelligence staff officer approves all products before dissemination. The intelligence staff develops and recommends priority intelligence requirements and information requirements for approval by the battalion commander and maintains the collection plan with the assistance of the analysis and control team. The intelligence staff directs all intelligence-collection operations involving battalion assets.

7-52. The intelligence staff identifies the need for collateral and SCI communications support and intelligence automated data processing systems support. The intelligence staff works with the signal staff in planning and coordinating SCI and automated data processing systems support. The intelligence staff coordinates GI&S requests and products, and conducts or coordinates for a wide variety of CI activities in support of OPSEC and operational area security. When a SOTF is established, the battalion intelligence staff officer serves in the operations center under the staff supervision of the operations center director. The intelligence staff officer is responsible for battalion information security, information systems security, personnel security, and special security programs. The intelligence staff officer exercises technical supervision over the battalion intelligence-training program to make sure it not only enhances military intelligence career management fields, but also includes the SF career management fields. When required, the battalion intelligence staff establishes a tactical SCIF adjacent to or in proximity of the operations center.

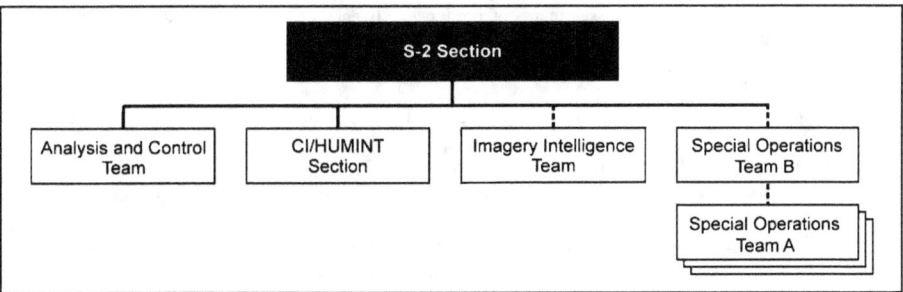

Figure 7-5. Battalion S-2 section

Analysis and Control Team

7-53. The analysis and control team is the hub of the S-2 section, and it has broad intelligence support responsibilities. The analysis and control team consists of all-source intelligence analysts and IMINT analysts. It responds to all-source intelligence support taskings from the battalion intelligence staff officer. The all-source intelligence analysts within the analysis and control team monitor the unit collection plan and identify new requirements to close intelligence gaps. The analysis and control team receives, validates, and prioritizes all requests for information from subordinate units and assigns them to the appropriate team of the intelligence staff. The analysis and control team informs the intelligence staff officer of priority intelligence requirement satisfaction and recommends new priority intelligence requirements, information requirements, and specific information requirements. It reports intelligence to higher or lateral units, based on guidance from the intelligence staff officer, and coordinates dissemination of intelligence to subordinate and attached units.

7-54. The analysis and control team is responsible for IPB, situation and target development, and battle damage assessment. All members of the analysis and control team contribute to the preparation of the intelligence estimate. The analysis and control team produces and updates target intelligence packages for the battalion. It maintains and uses the unit intelligence reference files and databases, including the SOCRATES, the JDISS-SOCRATES, and the CHIMS/CHATS.

7-55. Analysis and control team personnel translate all priority intelligence requirements from higher HQ into specific information requirements for tasking to subordinate units. Analysis and control team personnel prepare intelligence summaries, intelligence information reports, and responses to requests for information, and present intelligence briefings, as required. The analysis and control team provides all-source analysis and intelligence production support to the SF battalion and subordinate units in isolation conducting mission planning and preparation. The all-source intelligence analysts within the analysis and control team focus and refine the efforts of three single-source subordinate organizations—the IMINT team, the SIGINT-deployed TCAE, and the CI and HUMINT team.

7-56. Under the table of organization and equipment, the battalion MID has only one IMINT analyst assigned to it; however, he may be augmented with more IMINT analysts from the group MID or assisted by one or more all-source analysts. The IMINT analyst maintains comprehensive historical and current mission area imagery files in hard copy and digitized form. He uses the JDISS-SOCRATES or the EIW to provide imagery analysis support and production support. This support includes detailed terrain and facility analysis, annotated prints and target graphics, current imagery of targets and key terrain, battle damage assessment, and highly accurate imagery-based object measurement support. The IMINT analyst also develops reconnaissance requests for submission to higher HQ.

Counterintelligence/Human Intelligence Section

7-57. The primary function of the CI/HUMINT section is to collect information through interrogations, operational area security, OPSEC, and any other means identified. The CI/HUMINT section—

- Develops detailed assessments of foreign intelligence and security threats near deployed units and in the assigned area of operations.
- Recommends appropriate OPSEC and operational area security measures to reduce friendly vulnerabilities.
- Evaluates, if possible, the effectiveness of OPSEC measures implemented to correct identified friendly vulnerabilities.
- Briefs deploying teams on the latest threat data.
- Provides teams with technical advice and assistance to prepare them to establish and operate during UW missions.
- Supports operations by determining foreign vulnerabilities to deception.
- Provides the operations staff with recommendations for deception measures and evaluates their effectiveness through CI analysis.
- Requests external support, when necessary, to evaluate foreign reactions to friendly deception operations.
- Uses current intelligence automated data processing to exchange information with existing databases.

7-58. The SF battalion has two two-man HUMINT teams that may be task-organized, as required. The teams may be attached to a subordinate company or a joint interrogation facility. HUMINT team activities include—

- Interrogating EPWs and detainees.
- Debriefing detainees, returned U.S. personnel, and other persons of intelligence interest to the supported commander.
- Exploiting documents that appear to satisfy the supported commander's priority intelligence requirements directly.
- Conducting overt elicitation activities, such as liaison, escort, observer, and treaty verification missions.
- Conducting CI force protection source operations, as authorized in the theater CI plans.

Imagery Intelligence Team

7-59. Imagery analysts extract information and develop intelligence mainly from imagery collected from aerial and Tactical Exploitation of National Capabilities program sensors. A thorough knowledge of threat tactics and ground order of battle enables them to recognize, identify, locate, describe, and report information on objects, activities, and terrain from a variety of imagery products. Imagery analysts make associations between visible objects and configurations, and then analyze the results to determine strength, disposition, and enemy capabilities. The IMINT team—

- Analyzes imagery and reports specific information on threat operations, activities, dispositions, sustainment, communications, installations, and civilian activities, and their possible effect on operations.
- Prepares and maintains imagery prints to supplement and update maps for operational planning.
- Extracts information from imagery in direct support of unit mission area analysis.
- Supports combat assessment.

- Accomplishes imagery exploitation, reporting, and production for requirements from joint or theater imagery exploitation activities.
- Prepares and maintains an imagery database for the unit's specified mission areas.
- Serves as the primary collection manager at a battalion.

Special Operations Team B

7-60. The SF battalion special operations team B performs SIGINT and electronic warfare management functions, and provides control of SOT-As. Specific functions of the special operations team B include—
- Producing the SIGINT collection plan.
- Providing centralized operational control and collection tasking authority over deployed SOT-As.
- Analyzing and correlating intercepted SIGINT traffic from the SOT-As with data from other sources and then passing these products to the analysis and control team and TCAE or SIGINT processing centers.
- Providing technical support to SOT-As, as required.
- Publishing reports from SOT-As and nonorganic SIGINT systems into the national SIGINT system.

Special Operations Team A

7-61. The SOT-A is a low-level SIGINT collection team that intercepts and reports operational and technical information derived from tactical threat communications through prescribed communications paths. The mission of a SOT-A is to conduct SIGINT and electronic warfare in support of information operations to support Army special operations unit missions worldwide.

7-62. The primary roles of the SOT-A are electronic reconnaissance and protection, and support to foreign internal defense operations. Other SOT-A activities include signals research and target development, and support to personnel recovery missions. The SOT-A is assigned roles based upon an overall mission analysis and the commander's approval. Each battalion has three SOT-As. A SOT-A consists of the following three personnel:
- *Team sergeant.* The team sergeant is a sergeant first class and is the senior-ranking interceptor on the team; he performs many of the same duties as any supervising noncommissioned officer. The team sergeant performs voice intercept, processes highly sophisticated and complex radio transmissions, refines essential element of information requirements for identification and extraction, and supervises electronic support measures or electronic countermeasures of the team.
- *Second senior voice interceptor.* A staff sergeant holds the position of second senior voice interceptor on the team. He intercepts, identifies, and records designated foreign voice transmissions. He also operates equipment that collects and produces written records of nonstereotyped foreign voice radio transmissions. While collecting, he produces on-line activity records.
- *Voice interceptor.* The voice interceptor is a sergeant who intercepts, identifies, and records designated foreign voice transmissions. He also operates equipment that collects and produces written records of nonstereotyped foreign voice radio transmissions. While collecting, he simultaneously produces on-line activity records.

7-63. SOT-A members operate in remote and denied areas. In addition to their linguistic, international Morse code, and SIGINT skills, SOT-A personnel are cross-trained in critical tasks, tactical skills, and field craft.

7-64. The primary system for SOT-A electronic warfare support is the AN/PRD-14 Joint Threat Warning System Ground Signals Intelligence Kit. The Joint Threat Warning System Ground Signals Intelligence Kit is a lightweight, air-droppable electronic warfare support device capable of monitoring the radio frequency spectrum and conducting direction finding. Each SOT-A has one Joint Threat Warning System Ground Signals Intelligence Kit. Additionally, commercial off-the-shelf and Government off-the-shelf systems are available to assist SIGINT operations in frequency counting and in voice and data intercept. A majority of the items are procured through force modernization and the World Wide Web.

7-65. A SOT-A using organic handheld equipment can provide intercept and direction-finding of distress frequencies during personnel recovery missions. However, with the fielding of the combat survivor evader locator distress radio, this capability is reduced. Any support to personnel recovery must be assigned with specific reporting instructions and should be assigned only after a thorough mission analysis.

Company Intelligence Organization

7-66. The SF company operations warrant officer has staff responsibility for unit intelligence activities and training. At the operational detachment A level, the SF intelligence sergeant, career military occupational specialty 18F, supervises intelligence training, collection, analysis, production, and dissemination activities for his team. The intelligence sergeant assists the operations sergeant by preparing area studies, briefings, and combat orders. He helps conduct limited tactical questioning of EPWs. He briefs and debriefs patrols with or without the assistance of other detachment members.

7-67. In a UW or foreign internal defense operation, the SF intelligence sergeant may train a foreign intelligence section. Once deployed, he serves essentially as an intelligence staff officer. Therefore, he must maintain proficiency in the intelligence tasks, the MDMP, and handling and safeguarding classified material. He must also be a subject-matter expert on intelligence-collection systems and procedures, such as—

- Asymmetric software kit.
- Biometrics (Figure 7-6).
- Photography (film and digital).
- Communications.
- Protection.
- Tactical questioning and interrogation initial screening.
- Link analysis.
- Isolated personnel report preparation.
- Fingerprint identification system.

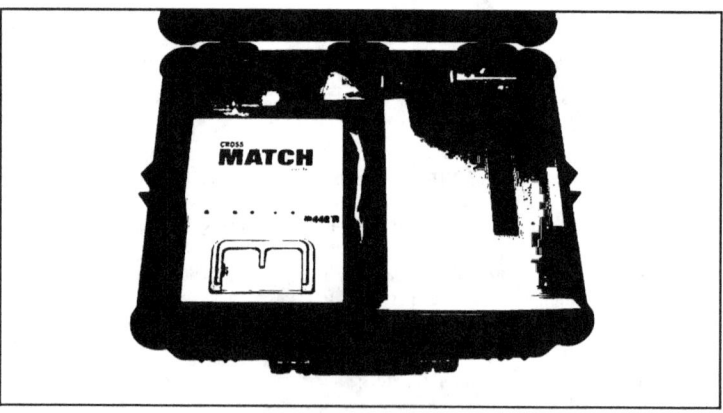

Figure 7-6. Cross Match biometrics kit

7-68. All SF intelligence sergeants must maintain a close relationship with the battalion intelligence staff and be able to perform all intelligence tasks. They are responsible for training their teams in intelligence tasks and recognize their team's intelligence requirements during isolation.

7-69. SF units at company and below may be required to work in an austere environment separated from a support mechanism. Decentralized operations do not relieve detachments from any duties, even intelligence analysis and production. Therefore, units may be augmented with intelligence personnel from the battalion or the group. Depending on the operational and intelligence requirements, the augmentation may be as simple as an additional CI or HUMINT collector, or as elaborate as one representative from each of the intelligence disciplines. The need for augmentation should be recognized and coordinated for as early in advance as possible to allow the battalion or the group to shift efforts and maintain seamless support to the remaining elements in the unit.

NONORGANIC INTELLIGENCE SUPPORT

7-70. SF units are consumers of products from all seven intelligence disciplines. By themselves, the disciplines and functions rarely provide a comprehensive picture of the threat or situational awareness. Instead, each provides fragments of information synthesized through an analytical process to approach total situational awareness. Different contingencies require tailored mixes of products from multiple disciplines to achieve this awareness. The following paragraphs, therefore, do not represent a hierarchy of usefulness.

OPEN-SOURCE INTELLIGENCE

7-71. OSINT is information of potential intelligence value available to the public. OSINT is essentially the research and monitoring of current events. It typically uses expertise outside the intelligence community—for example, at the national level, the Foreign Broadcast Information Service of the Open Source Center provides translations of foreign broadcast and print media. As information technology becomes more readily available, intelligence analysts must be able to exploit these sources, even at remote locations. OSINT may be the primary intelligence source for many operations. OSINT may also provide historical, cultural, economic, and environmental data from open-source material, media, and databases available. The ASK provides powerful tools to collect and manage open-source information.

HUMAN INTELLIGENCE

7-72. The role of HUMINT is to gather information from people and multiple media sources to identify enemy elements, intentions, composition, strength, dispositions, tactics, equipment, personnel, and capabilities. HUMINT relies on human sources and a variety of collection methods to gather information to satisfy the commander's intelligence requirements and cue other intelligence disciplines to collection opportunities. Organic CI agents, HUMINT collectors, and SFODA personnel collect and gather HUMINT information. Organic CI and HUMINT personnel can conduct CI force protection source operations when approved. HUMINT collection includes liaison activities, debriefing, tactical screening, interrogations, low-level source operations, and CI force protection source operations.

7-73. HUMINT is often the only source that can satisfy critical ARSOF intelligence requirements, whether from overt or controlled sources. Information collected by HUMINT means is often unavailable by technical means. HUMINT is particularly important during stability operations when the detection of emerging threats cannot be satisfied by other means. However, early planning, placement, and synchronization of assets are critical factors in the success of HUMINT operations. HUMINT operations normally require extensive lead times to train and rehearse, to gain access to a target area, and to begin the collection process. Though not the easiest collection means to employ, products may be critical to successful execution of special operations. HUMINT provides the ability to have eyes and ears on the ground. It assists in confirming intelligence from other sources. HUMINT provides situational awareness in the operational area and ground truth.

7-74. One of the imperatives for successful mission execution is sufficient intelligence, but obtaining adequate intelligence often requires that theater or national intelligence resources be applied at the expense of other needs. Using SOF reconnaissance and surveillance assets in denied or remote areas provides the potential for near-real-time information in the area of interest. SR operations involve a broad range of intelligence collection activities, including reconnaissance, surveillance, and target acquisition. The SR collection effort emphasizes joint or multinational intelligence requirements, not the requirements of an indigenous resistance organization. SR complements or augments national and theater collection systems (such as high-altitude IMINT or SIGINT) that are more vulnerable to weather, terrain, masking, and hostile countermeasures. Typical SR missions that support intelligence collection include—

- Initial contact with an indigenous resistance organization and an assessment of the resistance potential.
- Collection of strategic political, economic, psychological, or military information.
- Collection of critical military order of battle information—for example, weapons of mass destruction locations, types and capabilities, and intentions; commitment of second-echelon forces; and location of high-level HQ.
- Collection of technical military information.
- Target acquisition and surveillance of hostile mission command systems, troop concentrations, weapons of mass destruction, lines of communication, and other military targets of strategic or operational significance.
- Location and surveillance of facilities detaining hostages, prisoners of war, or political prisoners.
- Poststrike reconnaissance.
- Meteorological, geographic, or hydrographic reconnaissance to support specific air, land, or sea operations.

SIGNALS INTELLIGENCE

7-75. SIGINT includes all communications intelligence, ELINT, and foreign instrumentation SIGINT. The NSA exercises control over all SIGINT operations. SIGINT elements assigned to or supporting SF may exploit communications intelligence. Depending on the level required for subsequent analysis and reporting, units in the operational area may perform limited to in-depth analysis and forward collected information to higher levels for further processing. ELINT in support of SF may come from a number of sources, including assets attached to SF groups, national ELINT assets, or combatant command intelligence production centers. SF units use SIGINT products to—

- Prepare for and conduct infiltration and exfiltration.
- Locate actual or potential threat positions.
- Provide the deployed commander with immediate threat warning and situational awareness and information.
- Determine or analyze possible adversary COAs.

MEASUREMENT AND SIGNATURE INTELLIGENCE

7-76. MASINT is a highly sophisticated application of technology and processing techniques to detect and identify specific foreign weapons systems and threat activities on the basis of inadvertent signatures. This identification helps determine capabilities and intentions. One MASINT collection system organic to SF units is the improved remotely monitored battlefield sensor system, which can be employed to conduct SR and protection missions.

COUNTERINTELLIGENCE

7-77. CI incorporates aspects of counter-human intelligence, counter-imagery intelligence, and counter-signals intelligence to defeat or hinder threat intelligence collection and targeting. CI provides analysis of foreign intelligence threats, including espionage, sabotage, subversion, assassination, terrorism, and other threats.

7-78. CI counters or neutralizes threat and enemy intelligence collection efforts through collection, investigations, operations, analysis and production, and technical services. The role of CI is to detect, identify, track, exploit, and neutralize the multidisciplined intelligence activities of an expanding unknown number of global threats. CI is the key intelligence community activity supporting the mission to protect U.S. interests and equities.

7-79. CI elements contribute to situational understanding. They may corroborate intelligence from other intelligence disciplines, as well as cue other reconnaissance and surveillance assets. CI focuses on combating enemy and threat reconnaissance and surveillance activities that target Army personnel, plans, operations, activities, technologies, and other critical information and infrastructures.

7-80. CI operations are coordinated not only with the supported commander, but also between the Deputy Chief of Staff for Operations and Plans/operations staff officer or the Deputy Chief of Staff for Intelligence/intelligence staff officer and the G/S-2X. The G/S-2X provides CI (and HUMINT) collection expertise and advises the intelligence staff officer and the commander on all CI matters. The G/S-2X also participates in the reconnaissance and surveillance planning and synchronization efforts by recommending the best CI asset or combination of assets to satisfy the intelligence requirements. CI operations—
- Include specific actions that support the protection of the force.
- Counter the foreign multidisciplined intelligence threat.
- Counter foreign sabotage, subversion, assassination, and terrorism.
- Include the conduct of CI force protection source operations when approved by the theater special operations commander.

CI agents can support—
- Personnel security or information security.
- Physical security.
- OPSEC.

7-81. CI analysis provides commanders detailed assessments of foreign all-source intelligence and security threats near their operational bases and in their areas of operation. These foreign threat assessments are critical to the unit's OPSEC and base defense programs supporting the total effort. CI analysis supports deception operations by determining foreign intelligence collection assets and ways to thwart those collection efforts. CI analysis provides the operations staff recommendations of friendly activities to support the deception. If these activities are used, CI analysis helps evaluate their effectiveness.

GEOSPATIAL INTELLIGENCE

7-82. GEOINT consists of imagery, IMINT, and geospatial information. GEOINT products range from standard geospatial-data-derived products, such as two- and three-dimensional maps and imagery, to advanced products that incorporate data from multiple types of advanced sensor technology. All SF units are major consumers of GEOINT in all principal tasks. The NGA support officer at USASOC can assist the SF Command intelligence staff in obtaining specialized NGA products.

7-83. Imagery is a likeness or presentation of any natural or man-made feature or related object or activity and the positional data acquired at the same time the likeness or representation was acquired. Imagery includes products produced by space-based national intelligence reconnaissance systems and likenesses or presentations produced by satellites, airborne platforms, UASs, or other similar means. Imagery generally comes from three sources: national, civil, and commercial.

7-84. IMINT is derived from the exploitation of products from visual photography or infrared, electro-optical, and radar sensors. SF units use IMINT for target analysis, infiltration and exfiltration, and general reconnaissance or area orientation. The results of the exploitation and the annotated images may be incorporated into an all-source product focusing on a given enemy target, target type, or activity. Geospatial information provides the commander and staff with a fusion of weather and terrain, using current weather, digital or hard-copy maps, and input of changing terrain data from the field to obtain a common operational picture of the operational environment. IMINT products serve to—
- Identify potential enemy COAs.
- Assist in determining and selecting potential landing zones and drop zones.
- Identify high-value targets.
- Assist in selecting infiltration and exfiltration routes for rotary- and fixed-wing operations.
- Assist in assessments of staging bases.

7-85. Geospatial information is that information that identifies the geographic location and characteristics of natural or constructed features and boundaries on the earth, including statistical data; information derived from, among other things, remote sensing; mapping and surveying technologies; and mapping, charting, geodetic data, and related products.

TECHNICAL INTELLIGENCE

7-86. TECHINT, another multidisciplined function, focuses on foreign technological developments and the operational capabilities of foreign materiel that have practical application for military use. TECHINT products enable the exploitation of foreign weapons and other targetable systems. SF units often operate deep in hostile or denied territory; they are often first to discover, identify, and provide information on new or previously unidentified materiel.

OTHER NONORGANIC SUPPORT

7-87. SF groups and their subordinate elements are typically less reliant on higher-level intelligence support than other Army special operations units because of their more robust organic assets. Military intelligence Soldiers at all levels of an SF intelligence organization must be proficient in the procedures for obtaining intelligence from national, joint, and Service sources.

National-Level Support

7-88. National intelligence organizations conduct extensive collection, processing, analysis, and dissemination activities. National intelligence organizations routinely support JFCs while continuing to support national decisionmakers. However, the focus of these national organizations is not evenly split among intelligence customers and varies according to the situation and competing requirements. The SF intelligence staff should employ the extensive capabilities provided by these organizations whenever possible.

Theater-Level Support

7-89. Intelligence requirements for SF operations can be very generic and satisfied mainly by OSINT sources, or can be unique, be highly sensitive, and require compartmentalized handling. The Intelligence Operations Division of USASOC coordinates with USSOCOM and theater intelligence organizations to provide the intelligence that cannot be provided by subordinate units. The TSOC is primarily concerned with theater intelligence policy formulation, planning, and coordination. The TSOC intelligence staff—

- Ensures sufficient intelligence support is available for each mission tasked by or through the TSOC.
- Facilitates the theater intelligence organizations' collection, production, and dissemination of intelligence to meet Army special operations unit requirements.
- Coordinates joint special operations intelligence collection operations and the production and dissemination of target intelligence packages to support special operations targeting efforts.
- Tasks subordinate units to collect and report information that supports information requirements.
- Monitors the status of requests for information until the appropriate collection assets respond.
- Maintains adequate intelligence databases to support requirements.

SPECIAL FORCES SUPPORT TO THE INTELLIGENCE PROCESS

7-90. Every mission has inherent activities that satisfy the commander's critical information requirements for operational planning, protection, and area assessment. In some cases, no element, other than an SF operational detachment, may have access to collect this data. The use of advanced special operations supports the accomplishment of these tasks.

7-91. Trained and certified personnel use human sources to collect tactical and operational information regarding enemy intentions, capabilities, and order of battle on terrorists; sabotage, insurgent, and subversive activities; and any other activity influencing the protection of U.S. forces. These military source operations are comprised of activities using both contacts and recruited sources. Military source operations are conducted by trained personnel under the direction of the military commander. The gamut of HUMINT collection operations can be used.

7-92. The presence of advanced special operations trained Soldiers provides commanders with a HUMINT collection capability paralleling national-level intelligence organizations. Advanced special operations provide the ability to establish sources, penetrate networks, and then build networks of informants. Such operations require careful source synchronization with other agencies and units. Military source operations can provide definitive HUMINT on targets that may in turn be interdicted by Army special operations personnel or, by virtue of being a high-value or high-payoff target, targeted by robust conventional forces.

7-93. When approval is granted, the commander will designate an operational control element to coordinate, facilitate, and oversee the conduct of advanced special operations to support these operations. The operational control element may also be part of a military liaison element if tasked to conduct advanced special operations in support of operational preparation of the environment.

This page intentionally left blank.

Chapter 8
Intelligence Support to Special Operations Aviation

This chapter describes the United States Army Special Operations Aviation Command (USASOAC) missions and the intelligence units that support the command as well as the input that assigned Soldiers and aircrews provide to the intelligence process. The USASOAC units operate as part of a SOTF or a JSOTF. They give operating force commanders a means to infiltrate, resupply, and extract SOF. The units are not regionally focused as in the case of SF, CA, and MISO units. Instead, they operate in all theaters under almost any environmental condition. Specialized aircraft and crews operate with precise execution over extended ranges, under adverse weather conditions, and during times of limited visibility.

MISSION

8-1. The USASOAC includes a command HQ and a subordinate SOAR. The SOAR consists of a regimental headquarters and headquarters company and four battalions containing light attack, light assault, medium attack, airborne mission command, medium-lift, and heavy-assault helicopters and maintenance assets. The regiment's mission is to plan, conduct, and support special operations by clandestinely penetrating nonhostile, hostile, or denied airspace. The SOAR conducts missions across the range of military operations. The SOAR supports Army special operations units conducting joint, multinational, interagency, and liaison and coordination activities across the range of military operations. The SOAR organizes, supplies, trains, validates, sustains, and employs assigned aviation units for special operations missions.

INTELLIGENCE REQUIREMENTS

8-2. The regiment's operations have an inherent high degree of risk; successful execution relies on the intelligence staffs to gather, produce, and disseminate detailed aviation-specific intelligence to support planning and execution by staff and the executing crews. To successfully operate deep within hostile territory, aircrews must avoid enemy detection. Avoiding enemy acquisition systems, therefore, is critical. Current intelligence on the location, status, and operating modes and frequencies of enemy acquisition and tracking systems is essential. The SOAR uses intelligence to plan routes and determine the needs and settings of aircraft survivability equipment. It also uses intelligence to determine external support requirements. Mission planners use combat information and intelligence to plot infiltration and exfiltration routes and to recommend landing zones. Both general military and aviation-specific intelligence is vital to SOA planners developing input to a plan or special operations mission planning folder. Typical required planning information includes the following:
- Air order of battle.
- Air defense order of battle.
- Ground order of battle.
- Electronic order of battle.
- Naval order of battle.
- Imagery.
- Maps.

Chapter 8

- Climatology and terrain analysis.
- Recent threat activity and actions.

INTELLIGENCE ORGANIZATION

8-3. The only organic intelligence support in the SOAR and its subordinate battalions is their respective intelligence staffs. The organization and functions of the regiment and battalion intelligence staffs are the same, with the exception of the intelligence operations and security sections found at regimental level. The configurations of the regimental and battalion intelligence sections are depicted in Figure 8-1 and Figure 8-2, page 8-3. The regiment's intelligence section is staffed by all-source, imagery, geospatial, signals, and CI analysts; a special security officer; and an intelligence operations specialist. The regimental intelligence staff can augment battalion intelligence sections when required.

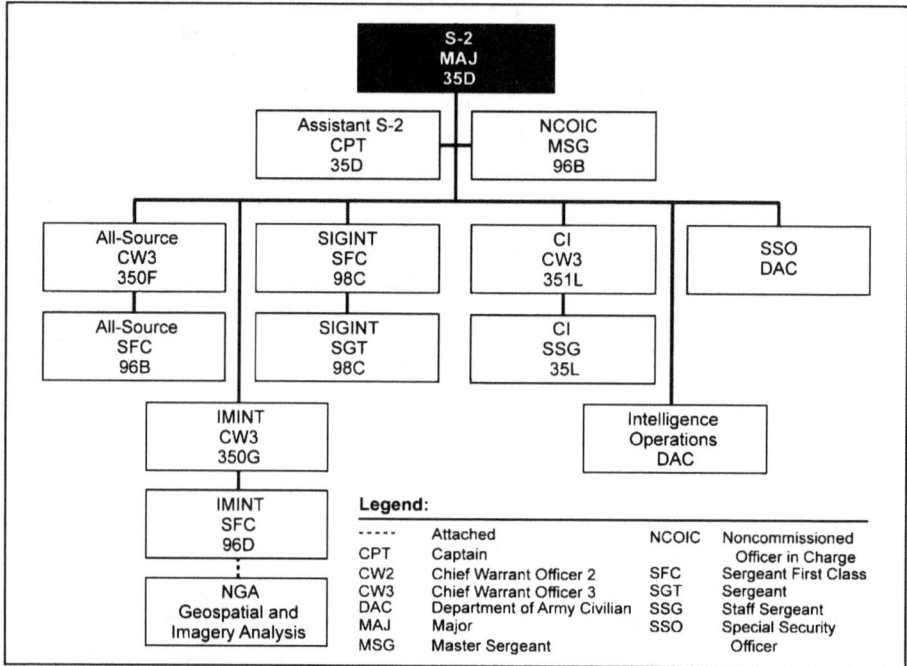

Figure 8-1. Intelligence organization of the Special Operations Aviation Regiment

INTELLIGENCE OFFICER RESPONSIBILITIES

8-4. The regimental intelligence officer is a military intelligence major who is the primary staff officer responsible for all aspects of intelligence, CI, and security support in garrison. He coordinates weather support through attached USAF teams, manages the regiment map warehouse, provides macro intelligence for future operations, coordinates intelligence personnel augmentation for deployed units, and oversees CI support. The regimental intelligence officer pursues emerging technologies and coordinates required systems, organizations, and connectivity issues with USASOC. He manages the SCIF through the special security officer and ensures efficiency and security of classified message traffic.

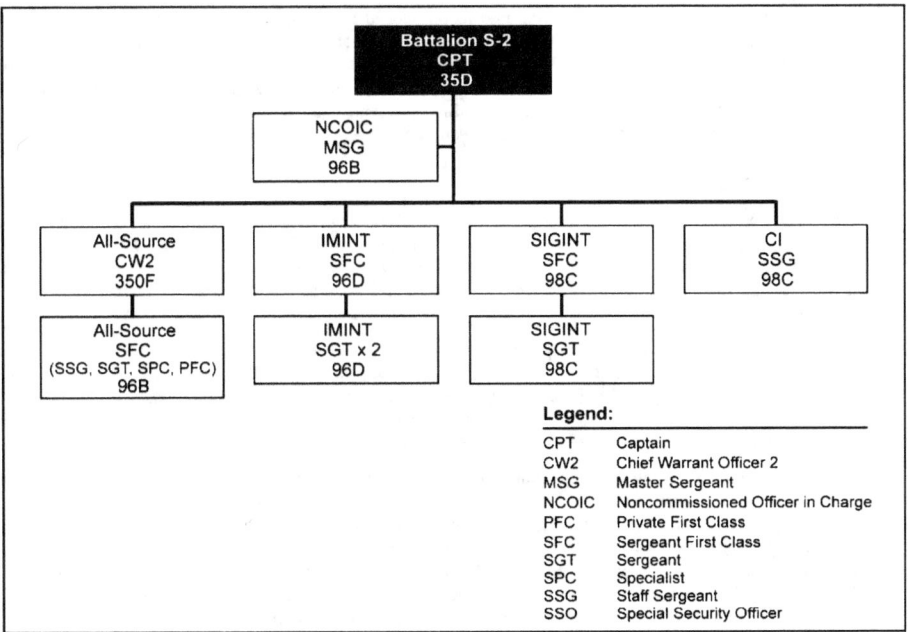

Figure 8-2. Intelligence organization of the Special Operations Aviation Regiment battalion

ALL-SOURCE SECTION

8-5. The all-source analysts analyze intelligence using products from all intelligence disciplines. They study, analyze, integrate, and process threat information on targets identified in theater planning documents or for contingency operations in support of missions or training. The analysts track upgrades to specified country ground and electronic order of battle in addition to the modification and proliferation of surface-to-air systems. The analysts can augment battalion intelligence sections, as required. They are led by an all-source intelligence technician and a senior intelligence analyst.

8-6. The all-source analysts' primary task is the identification and tracking of global threats for potential future missions using all available intelligence disciplines. Their focus is the modification and proliferation of surface-to-air missiles. They archive products, including IPB overlays, situation updates, intelligence assessments, and weapons data research. The all-source technician coordinates with the regiment's Systems Integration and Management Office for the identification, research, and sharing of threat systems in relation to current aircraft survivability equipment capabilities. He also interfaces with USASOC G-2 and G-8 with regard to new or updated analytical technology. He then coordinates the fielding of new analytical technology in the regiment and its associated training for the all-source analysts.

SIGNALS INTELLIGENCE SECTION

8-7. The signals analysts are responsible for providing SIGINT and ELINT analysis to aircrews and mission planners to ensure their aircraft flights are safe and undetected. These analysts are led by a senior SIGINT analyst. The analysts can augment battalion intelligence sections when required.

8-8. The analysts provide near-real-time adversary communications intelligence and ELINT situational awareness. They focus on premission analysis of both route threat and electronic detection capabilities of target countries. The SIGINT section is able to produce the most current electronic order of battle and operations/activity clocks on associated electronic order of battle for the assigned area of operations and area of interest. The analysts will help determine current locations of electronic and communications equipment, and trends and patterns of equipment and users. The senior SIGINT analyst is responsible for researching, recommending, acquiring, and fielding new and updated technology for the SOAR through the Systems Integration and Management Office and USASOC G-2 or G-8.

IMAGERY INTELLIGENCE SECTION

8-9. The regiment's IMINT section is the center of the regiment's garrison imagery operations and supports forward-deployed units. The analysts are led by an IMINT technician. There are also two NGA personnel (one imagery analyst and one geospatial analyst) who provide a conduit to the NGA for acquisition of special GI&S products. When required, the regiment imagery analysts may be tasked to augment the battalions.

8-10. The NGA has attached a geospatial and imagery analyst to the SOAR to provide a direct link to NGA products and support. The geospatial analyst provides mentorship to the imagery analysts, instructs them on the use of geospatial software, and generates specific products to support the unit mission. The NGA imagery analyst provides mentorship to the other imagery analysts and instruction on use of imagery software and analysis. The NGA geospatial and imagery analysts are capable of deploying with the unit.

8-11. The regiment's imagery technician is responsible for pursuing emerging imagery technology and program upgrades for the regiment. When technology is fielded or updated, the IMINT technician is responsible for researching, planning, and coordinating training for the imagery analysts. To support deployed units, the imagery analysts order, receive, manipulate, annotate, and disseminate IMINT products (full frame and chipped imagery), both printed and electronic copies. These products include secondary imagery products, helicopter landing zones and drop zones, noncombatant evacuation support, route assessments, terrain analysis, ingress/egress route analysis, line-of-sight studies, and point targets such as embassies, ports, airfields, and training areas.

INTELLIGENCE OPERATIONS SECTION

8-12. The regiment's intelligence operations section performs a full range of administrative, intelligence, and intelligence-related security functions. The section plans, develops, coordinates, and evaluates intelligence and security program goals and objectives and assists with their implementation. Intelligence specialties include all-source operations, collection management, and intelligence system requirements. The intelligence operations specialist is responsible for performing the following functions:

- Develops all-source systems requirements:
  - Monitors the intelligence futures program and Tactical Exploitation of National Capabilities program in order to leverage developing techniques and technologies to increase intelligence productivity.
  - Identifies shortcomings in unit intelligence architecture and equipment capabilities.
  - Provides expertise, direction, and information regarding the acquisition of new intelligence systems.
- Oversees or otherwise conducts the duties of collection requirements, operations, and production management:
  - Generates, processes, and validates both ad hoc and standing all-source intelligence collection requirements.
  - Identifies and registers unit intelligence requirements to the DIA and manages the DIA accounts for the regiment.
- Serves as the unit's DOD Intelligence Information System site information system security manager with overall responsibility for the DOD Intelligence Information System information system security program:
  - Drafts appropriate facility standing operating procedures and chairs the site's DOD Intelligence Information System configuration control board.

- Has overall responsibility for coordinating and controlling access to worldwide intelligence databases.
- Establishes, maintains direct liaison, and coordinates unit intelligence and security requirements with USASOC; national and theater-level intelligence organizations; the military Services; combatant, geographic, and functional commands; and other appropriate agencies.

COUNTERINTELLIGENCE SECTION

8-13. The regiment's CI personnel detect, evaluate, counteract, or prevent foreign intelligence collection, subversion, sabotage, and terrorism. They determine security vulnerabilities and recommend countermeasures. The CI section supports decisive, shaping, and sustaining operations, and CI support to security for all training exercises and operational deployments. The CI section is led by a CI technician and includes a CI agent. Regimental CI personnel may augment the battalions, as required.

8-14. CI assets primarily focus on warfighting protection variables for deployed regiment elements. They conduct liaison with U.S. and HN intelligence and law enforcement agencies, as required. In response to situation and collection orders, CI personnel plan, coordinate, and conduct CI liaison, overt collection, and displaced civilian debriefings. They recommend CI threat countermeasures, ensure compliance with intelligence oversight regulations, and continually assess the effectiveness of the base OPSEC countermeasures and base security plans. They also support the OPSEC plan by providing TARP briefings and limited investigations.

SPECIAL OPERATIONS WEATHER TEAM

8-15. The SOWT is responsible for accurate, detailed, and specific weather and environmental data. Determining potential effects of weather is a critical aspect of mission planning. Severe weather conditions can degrade flight and target acquisition capabilities. Atmospheric conditions affect the propagation of aircraft noises, which impact route selection. Even solar activity can degrade sensitive communication equipment. SOWTs provide timely weather support for the regiment. Direct weather support includes, but is not limited to—

- Forecasts of general weather conditions and specific meteorological data elements as described in the 24-hour forecast.
- Geophysical information and climatic studies and analyses.
- Weather advisories, warnings, and specialized weather briefings, to include flight weather briefings of routes and objective areas.
- Lunar and solar illumination data, and light angle of incidence.
- Expected temperature variations along flight routes.
- Atmospheric factors (ceilings, visibility, pressure, and wind conditions).
- Hazards to aircraft (icing and turbulence).
- Weather effects on timelines and schedules.
- Tidal data and water temperatures.

NONORGANIC INTELLIGENCE SUPPORT

8-16. Since organic intelligence support is limited, intelligence staffs must be adept at leveraging support from the intelligence community. The SOAR normally operates as part of a task force. Consequently, unit intelligence staffs should integrate their intelligence operations with those of other task force units—especially when those units have more robust intelligence support structure.

COUNTERINTELLIGENCE/HUMAN INTELLIGENCE SUPPORT

8-17. Although the SOAR has its own organic CI section, it may require augmentation from personnel assigned to higher echelons or liaison with CI personnel assigned to adjacent units. Support from nonorganic assets includes hazard reconnaissance, vulnerability assessments, and liaison support. The SOAR does not have any organic HUMINT capability; intelligence sections must request national or task force HUMINT assets for source operations, interrogation, liaison, and document exploitation. The regiment's CI personnel can perform tactical questioning and debriefings with nonorganic linguistic support (Appendix D) and proper training. Appendix E discusses document exploitation and handling.

IMAGERY INTELLIGENCE SUPPORT

8-18. The SOAR requires immediate access to current, detailed imagery products to plan all aspects of their precision assault missions. In addition to its organic IMINT support, joint and national assets normally provide tactical support to the regiment's operations. IMINT sources include UAS feed and radar, and infrared and electro-optic imagery. Planners use this detailed imagery to develop sophisticated target folders. IMINT enables effective planning for missions and contingencies, such as determining clearance distances for landing zones and locating hazards, such as wires or tall obstacles.

SIGNALS INTELLIGENCE SUPPORT

8-19. The SOAR battalions do not have an organic SIGINT capability. When regiment support is not available, they rely on joint and national ELINT databases and collectors to locate and identify electronic warfare emitters in the operational area that could impact flight routes. Locating these emitters and their associated weapons systems is crucial to ensuring that the aircraft are able to fly undetected to and from their objectives.

SPECIAL OPERATIONS AVIATION REGIMENT SUPPORT TO THE INTELLIGENCE PROCESS

8-20. SOAR units gather specific operational information as part of their overall mission. In addition to this information, the intelligence staff researches current intelligence databases and refines applicable reporting into aviation-specific intelligence products. These products are then posted to intelligence databases as a resource for aviation or aviation-supported units.

Appendix A
Open-Source Intelligence and Information

Army special operations units have traditionally gathered much of the information and resultant intelligence needed for operations from open sources. In addition, if properly channeled into the intelligence process, open-source information gathered by units in the routine conduct of special operations missions can provide, through application of the intelligence process, timely intelligence to the entire force. Frequently, Army special operations units gather specialized information that cannot readily be gathered by conventional units with their organic resources.

OSINT operations are integral to the intelligence process. Publicly available information typically forms a baseline for all intelligence operations and intelligence products. The availability, depth, and range of publicly available information enable intelligence organizations to satisfy many intelligence requirements without the use of specialized human or technical means of collection. This appendix introduces the fundamentals of open-source information and intelligence as they apply to Army special operations units and missions.

DEFINITIONS AND TERMS

A-1. *Open-source intelligence* is information of potential intelligence value that is available to the general public (JP 1-02, *Department of Defense Dictionary of Military and Associated Terms*). The National Defense Authorization Act described open-source intelligence as intelligence products produced from publicly available information that is collected, exploited, and disseminated in a timely manner to an appropriate audience for the purpose of addressing a special intelligence requirement. Two important terms associated with the concept of open-source information are—

- *Open source*: Any person, group, or system that provides information without the expectation of privacy—the information and the relationship, or both, are not protected against public disclosure.
- *Publicly available information*: Data, facts, instructions, or other material published or broadcast for general public consumption; available on request to a member of the general public; lawfully seen or heard by any casual observer; or made available at a meeting open to the general public.

A-2. Open sources broadcast, publish, or otherwise distribute unclassified information for public use. The collection means (techniques) for gathering publicly available information from these media of communications are overt and unobtrusive. Other intelligence disciplines use clandestine or intrusive techniques to collect private information from confidential sources. Confidential sources and private information are defined as follows:

- *Confidential source*: Any person, group, or system that provides information with the expectation that the information, relationship, or both, are protected against public disclosure.
- *Private information*: Data, facts, instructions, or other material intended for or restricted to a particular person, group, or organization. There are two subcategories of private information: classified information and controlled unclassified information.

A-3. Classified information requires protection against unauthorized disclosure and is marked to indicate its classified status when in documentary or readable form. Controlled unclassified information requires the application of controls and protective measures (for example, sensitive but unclassified, or for official use

Appendix A

only), not to include those that qualify for formal classification. Open sources and publicly available information may include, but are not limited to—

- *Academia.* Courseware, dissertations, lectures, presentations, research papers, and studies in both hard copy and soft copy obtained from any institution of education.
- *Governmental and nongovernmental organizations.* Databases, posted information, printed reports, and sometimes propaganda produced by organizations.
- *Commercial and public information services.* Broadcast, posted, and printed news.
- *Libraries and research centers.* Printed documents and digital databases on a range of topics, as well as knowledge and skills in information retrieval.
- *Individuals and groups.* Handwritten, painted, posted, printed, and broadcast information (for example, art, graffiti, leaflets, posters, and Web sites).

A-4. The primary media that open sources use to communicate information to the general public are—
- Public speaking forums.
- Public documents.
- Public broadcasts.
- Internet sites.

A-5. Public speaking, the oldest medium, is the oral distribution of information to audiences during events that are open to the public or occur in public areas. These events or forums include, but are not limited to, academic debates, educational lectures, news conferences, political rallies, public government meetings, religious sermons, and science and technology exhibitions. Neither the speaker nor the audience has the expectation of privacy when participating in a public speaking forum. Unlike other open-source collection, monitoring public speaking events is done through direct observation and, due to its overt nature, could entail risk to the collector.

A-6. A document is any recorded information regardless of its physical form or characteristics. Like public speaking, public documents have always been a source of intelligence. Documents provide in-depth information about the operational environment that underpin the ability to plan, prepare for, and execute military operations. During operations, documents such as newspapers and magazines provide insights into the effectiveness of information-related activities conducted to influence. Books, leaflets, magazines, maps, manuals, marketing brochures, newspapers, photographs, public property records, and other forms of recorded information continue to yield information of intelligence value about operational environments. Sustained document collection contributes to the development of studies about potential operational environments. Collection of documents on the operational and technical characteristics of foreign materiel aids in the development of improved U.S. tactics, countermeasures, and equipment.

A-7. A public broadcast entails the simultaneous transmission of data or information for general public consumption to all receivers or terminals within a computer, radio, or television network. Public broadcasts are important sources of current information about the operational environment. Television news broadcasts often provide the first indications and warnings of situations that may require the use of U.S. forces. Broadcast news and announcements enable personnel to monitor conditions and take appropriate action when conditions change within the operational area. News, commentary, and analysis on radio and television also provide windows into how governments, civilians, news organizations, and other elements of society perceive the United States and U.S. military operations.

A-8. Internet sites enable users to participate in a publicly accessible communications network that connects computers, computer networks, and organizational computer facilities around the world. The Internet is more than just a research tool. It is a reconnaissance and surveillance tool that enables intelligence personnel to locate and observe open sources of information.

A-9. Through the Internet, trained collectors can detect and monitor Internet sites that may provide indications and warnings of enemy intentions, capabilities, and activities. Collectors can monitor newspaper, radio, and television Web sites that support information-related capability assessments. Collectors can conduct periodic searches of Web pages and databases for content on military order of battle, personalities, and equipment. Collecting Web page content and links can provide useful information about relationships between individuals and organizations. Properly focused, collecting and processing publicly available information from Internet sites can help analysts and decisionmakers understand the operational environment. Figure A-1 lists Internet components and elements.

Components	Elements
Communications	Chat
	E-mail
	News
	Newsgroup
	Webcam
	Webcast
	Weblog
Databases	Commerce
	Education
	Government
	Military
	Organizations
Information (Web Page Content)	Commerce
	Education
	Government
	Military
	Organizations
Services	Dictionary
	Directory
	Downloads
	Financial
	Geospatial
	Search
	Technical Support
	Translation
	Uniform Resource Locator (URL) Lookup

Figure A-1. Internet components and elements

CONSIDERATIONS

A-10. OSINT operations must comply with Army Regulation (AR) 381-10, *U.S. Army Intelligence Activities*, and Executive Order 12333, *United States Intelligence Activities*, on the collection, retention, and dissemination of information on U.S. persons, both within and outside of the United States. Information-gathering activities by personnel outside the military intelligence occupational specialty must adhere to these same guidelines. OSINT organizations can be overwhelmed by the volume of information to process and analyze. OSINT operations require qualified linguists to collect and process non-English-language information. In addition to these common considerations, personnel responsible for planning or executing OSINT operations must consider the following:

- *Legal restrictions.* Intelligence organizations whose principal missions are CI, HUMINT, and SIGINT must comply with applicable DOD directives and regulations that govern contact with and collection of information from open sources. For example, Department of Defense Directive (DODD) 5100.20, *National Security Agency/Central Security Service (NSA/CSS)*, prohibits SIGINT organizations from collecting and processing information from public broadcasts with the exception of processing encrypted or "hidden meaning" passages.

Appendix A

- *Operations security.* More than any other intelligence discipline, the OSINT discipline could unintentionally provide indicators of U.S. military operations. Information generally available to the public, as well as certain detectable activities such as open-source research and collection, can reveal the existence of, and sometimes details about, classified or sensitive information or undertakings. Purchasing documents, searching an Internet site, or asking questions at public events are examples of detectable open-source research and collection techniques that could provide indicators of U.S. plans and operations.
- *Deconfliction.* Intelligence and operations staffs deconflict reconnaissance and surveillance operations during planning to avoid compromising other intelligence discipline operations.
- *Deception and bias.* Deception and bias are of particular concern in OSINT operations. Unlike other disciplines, OSINT operations do not normally collect information by direct observation of activities and conditions within the area of interest. OSINT operations rely on secondary sources to collect and distribute information that the sources may not have observed themselves. Secondary sources such as government press offices, commercial news organizations, nongovernmental organization spokesmen, and other information providers can intentionally or unintentionally add, delete, modify, or otherwise filter the information they make available to the general public. These sources may also convey one message in English for U.S. or international consumption and a different non-English message for local or regional consumption.
- *Intellectual property.* AR 27-60, *Intellectual Property*, prescribes policy and procedures with regard to the acquisition, protection, transfer, and use of patents, copyrights, trademarks, and other intellectual property by the Department of the Army. It is Army policy to recognize the rights of copyright owners consistent with the Army's unique mission and worldwide commitments. As a general rule, Army organizations will not reproduce or distribute copyrighted works without the permission of the copyright owner unless such use is within an exception under U.S. Copyright Law or required to meet an immediate, mission-essential need for which alternatives are either unavailable or unsatisfactory. According to the U.S. Copyright Office, "fair use" of a copyrighted work for purposes such as criticism, comment, news reporting, teaching, scholarship, or research, is not an infringement of copyright.

A-11. Prior to conducting open-source information and/or intelligence gathering, hardware or software may need to be acquired. Purchase of additional audio or video equipment or adapting software or hardware for organic equipment may be required. Additional software for data storage, computer security, applications for saving Web content/metadata, and translation may be required. Purchase of information in various forms, such as books and videos as well as subscriptions to periodicals and information services, may be necessary as well. In addition, types of open source may require other specific actions such as—

- *Public speaking forums.* Collecting information at public speaking forums requires close coordination to ensure the overt collection is integrated with the information operations annex to the operation plan (OPLAN), is lawful, and is synchronized with HUMINT activities. The information collection plan describes how the unit will—
    - Coordinate with the public affairs officer, the Staff Judge Advocate (SJA), and the S-2X prior to surveillance of public speaking forums.
    - Identify additional personnel and equipment to support protecting the warfighting functions.
- *Public documents.* Prior to deployments, Army special operations units rely on nonorganic assets to obtain information in documents available outside the area of responsibility. While deployed, Army special operations units may obtain exploitable documents with organic resources. Army special operations units—
    - Coordinate document collection, processing, and production activities with those of joint, interagency, and multinational organizations and the National Media Exploitation Center.
    - Request and integrate document exploitation support from higher echelons.
- *Public broadcasts.* The Open Source Center, covered in detail below, collects, processes, and reports international and regional broadcasts. Doing so enables deployed organizations to use

their resources to collect and process information from local broadcasts that are of interest to the command or only accessible from within the operational area. Army special operations units—
- Coordinate broadcast collection, processing, and production activities with those of the Open Source Center.
- Use Internet collection and processing resources to collect the broadcast from Webcasts on the radio and television station's Internet site.

• *Internet.* Echelons above corps organizations such as the INSCOM's Asian Studies Detachment and the Joint Reserve Intelligence Center conduct sustained Internet collection, processing, and intelligence production. Army special operations units and organizations routinely monitor Internet sites of relevance to specific regions, countries, and locales. Internet collection, processing, and analysis are coordinated with those of joint, interagency, and multinational organizations.

OPEN SOURCE CENTER PRODUCTS AND TOOLS

A-12. In response to language in the Intelligence Reform and Prevention of Terrorism Act of 2004 and recommendations in the Silberman-Robb Commission calling for more effective use of open sources to support intelligence, the newly established Director of National Intelligence created the Director of National Intelligence Open Source Center at the CIA on 1 November 2005. The Director of National Intelligence assigned the Director, CIA, as Executive Agent for this center and directed that the center be built on the capabilities and expertise of CIA's Foreign Broadcast Information Service and report directly to the Director, CIA, in implementing Director of National Intelligence strategy and guidance.

MISSION

A-13. The Open Source Center executes Director of National Intelligence strategy and guidance, and supports Director of National Intelligence nurturing of distributed open-source architecture across the intelligence community. The Open Source Center manages that distributed enterprise through an inclusive approach to networking expertise and capabilities that exist not only within the intelligence community, but also across the government and throughout the private sector and academia. The Open Source Center works with the combatant commands and the entire DOD, down to the small-unit level, to deconflict strategies and minimize redundancy. In conjunction with colleagues across the open-source community, the Open Source Center develops and provides a number of centralized services that include specialized training, the building of common procedures and policies related to such issues as copyright and use of the Internet, and content procurement.

FOREIGN BROADCAST PRODUCTS AND SERVICES

A-14. The Open Source Center maintains a worldwide network of multilingual regional experts. They respond to intelligence requirements using open sources, including radio, television, newspapers, news agencies, databases, and the Internet. The Open Source Center monitors open sources in more than 160 countries in over 80 languages and acquires open-source data worldwide for organizations across the military and government, down to local law enforcement.

A-15. Open Source Center products and services include—
- *Analysis.* Open Source Center analyses range from short, time-sensitive products that explain the media treatment of issues on the U.S. Government policy agenda to longer analytic pieces that examine issues or the content and behavior of a set of media over time to detect trends, patterns, and changes related to U.S. national security interests.
- *Media guides.* Media guides offer a comprehensive characterization of the media of a country or region and provide an overview or characterization of the larger media environment, including what makes up the media of a country, how those media operate, who uses the media and how they use it, and other factors, such as literacy rates, press laws, economic status, and demographics that affect the media and their behavior or use.

Appendix A

- *Commentator profile.* Commentator profiles provide detailed information on one or more media personalities in a particular country, outlining their influence, background, views, and biases on key topics. The focus is on personalities who speak or write about issues of importance to the United States or who have influence with their government, businesses, or large segments of the general population.
- *Open Source Center reports.* Through its worldwide access to foreign media and other publicly available material, the Open Source Center provides translations and transcriptions of the latest political, military, economic, and technical information gleaned from foreign open sources. Reports are derived from select sources, such as jihadist Web sites or daily editorials, on topics related to the national intelligence priorities.
- *Video services.* Video Services Division collects, analyzes, and disseminates more than 550 channels of foreign and domestic television and Internet video 24 hours a day, 7 days a week. Video Services Division can access between 16,000 and 34,000 additional channels. Video Services Division has a library spanning 45 years.
- *Internet exploitation tools.* Open Source Center's Internet Exploitation Team provides training in, as well as access to, a large data repository, including an Internet cache and other purchased content, and a suite of data mining tools to allow research of Web-based and other open sources.
- *Opensource.gov content management services.* Opensource.gov content management capabilities include highlighting individual open-source products on the site. These content management capabilities can be granted to government employees or contractors, and require minimal training.
- *Product creation tools and systems.* For complex work processes, particularly those involving a distributed workforce or a multistep workflow, the Open Source Center's Processing in a Collaborative Environment system provides Web interfaces to enable product creation, editing, and dissemination. Users can direct items for translation, editing, and dissemination from any point on the globe with Internet access.

INTERNET SOURCES

A-16. Although not the only source of OSINT, vast amounts of information that Army special operations units can process into OSINT are available on the Internet. All units can use the Internet for some of their information needs. Certain units—for instance, MISO and CA—may find the majority of their initial information requirements on the Internet. The nature of the Internet lends itself to two types of unreliable information. The first is deception and disinformation, which is deliberate distortion of the truth or of facts. This type of unreliable information can be characterized as malicious. Misinformation is that class of information that is unintentionally inaccurate or incomplete. Large amounts of information on the Internet are posted by single individuals or small groups that may or may not vet their information or subject it to any sort of scholarly review. This lack of review can result in errors, incompleteness, shallowness, or superficiality of the information. This sort of information is rarely malicious but rather indicative of the freedom that exists to post any content on the Internet. The problem with this sort of information is that it may propagate itself from Web site to Web site as more and more sites with unproven content pick up on an interesting or entertaining article. The following paragraphs detail the advantages, disadvantages, and potential problems with exploiting Internet sources.

THE INVISIBLE WEB

A-17. Publicly available information also resides in many databases on the Internet. This information cannot, however, be accessed using a commercial search engine. This information is considered "invisible" or "deep web" because it resides on a Web site designed around a database; there are no static pages to index (Table A-1, page A-7). Some commercial vendors create fee-based databases with public information. Some of these commercial services are accessible through Army Knowledge Online (AKO) or the Open Source Center (Table A-2, page A-7). Operators of fee-based data services have an economic interest in the scope and dissemination of the data they offer. Improper use or dissemination could be infringement on the copyright law protections of the owner of the database. Army special operations units gain the opinion of the supporting Judge Advocate General office before disseminating information from a fee-based data service.

Table A-1. Invisible Web databases

Information Need	Location
Scholarly Documents Across Disciplines	Google Scholar: Google Scholar searches specifically for scholarly literature, including peer-reviewed papers, theses, books, preprints, abstracts, and technical reports from all broad areas of research. Public Internet at http://scholar.google.com/.
Lessons Learned: Intelligence	Center for Army Lessons Learned (CALL): Public Internet at http://call.army.mil/ or use AKO password for For Official Use Only (FOUO) information.
	Intelligence Center Online Network: Observations, Insights, and Lessons Learned. Public Internet with AKO password at http://iconportal.hua.army.mil.
Scientific and Technical Information	Defense Technical Information Center (DTIC): Public Internet with password at http://www.dtic.mil.
	DTIC Research and Engineering Portal: Public Internet with password at https://rdte.osd.mil.
	Scientific and Technical Information Network: Public Internet with password at https://dtic-stinet.dtic.mil/. This site is more restricted than DTIC.
Intelligence, Emerging Threats, Defense, and Military Information	World Basic Information Library: Research library located on OSIS and managed by the Foreign Military Studies Office.
Worldwide Infrastructure	Intelligence Road/Rail Information System: This is a Web-based portal to worldwide infrastructure and real-time data. Public Internet with password at https://www.irris.tea.army.mil/irris/site/.

Table A-2. Military-accessible fee-based databases

Information Need	Army Knowledge Online http://us.army.mil	Open-Source Information System http://www.osis.gov
Index for Full-Text Journal, Magazine, and Newspaper Articles	Academic Search Premier MasterFile Premier	Academic Search Premier
Defense and Military Information	Periscope Military and Government Collection	Jane's Online
Country Studies	Country Watch Publications via Military and Government Collection	Oxford Analytica China Vitae
Terrorism	Periscope	Jane's Online
Global Analysis and Events	Oxford Analytica Jane's Online	
World Economic and Business Indicators	The Economist Oxford Analytica The Economist Intelligence Unit (EIU Data Services) http://www.cosp.osis.gov/pages/eiu.htm	

INTERNET SOURCE RELIABILITY

A-18. Establishing the reliability of Internet sources focuses on determining the nature of the person or organization producing the site rather than the content of the Web pages. Joint doctrine establishes a reliability rating code in JP 2-01. This reliability rating structure is useful to categorize Web site producers. Reliability ratings range from A (Reliable) to F (Cannot Be Judged) as shown in Table A-3, page A-8. If the source is new, they rate the source as F (Cannot Be Judged). An F rating does not necessarily mean the source is unreliable, but that the collection and processing personnel have no previous experience with the

Appendix A

source upon which to base a determination. The reliability of all sources must be determined to include if it remains indeterminate. Unreliable sources may have usefulness in some searches in drawing negative inferences—for example, a search on propaganda—but only if thoroughly documented as coming from an unreliable source.

A-19. A modified scale (Table A-4) of information credibility serves to further classify open sources in general and Internet sites in particular. This scale runs from confirmed credibility (1) to "cannot be judged" (8). It also includes codes for misinformation and deliberate deception.

Table A-3. Source reliability

Code	Rating	Description
A	Reliable	No doubt of authenticity, trustworthiness, or competency; has a history of complete reliability; usually demonstrates adherence to known professional standards and verification processes.
B	Usually Reliable	Minor doubt about authenticity, trustworthiness, or competency; has a history of valid information most of the time; may not have a history of adherence to professionally accepted standards but generally identifies what is known about sources feeding any broadcast.
C	Fairly Reliable	Doubt of authenticity, trustworthiness, or competency but has provided valid information in the past.
D	Not Usually Reliable	Significant doubt about authenticity, trustworthiness, or competency but has provided valid information in the past.
E	Unreliable	Lacking in authenticity, trustworthiness, and competency; history of invalid information.
F	Cannot Be Judged	No basis exists for evaluating the reliability of the source; new information source.

Table A-4. Open-source information credibility

Code	Rating	Description
1	Confirmed	Confirmed by other independent sources; logical in itself; consistent with other information on the subject.
2	Probably True	Not confirmed; logical in itself; consistent with other information on the subject.
3	Possibly True	Not confirmed; reasonably logical in itself; agrees with some other information on the subject.
4	Doubtfully True	Not confirmed; possible but not logical; no other information on the subject.
5	Improbable	Not confirmed; not logical in itself; contradicted by other information on the subject.
6	Misinformation	Unintentionally false; not logical in itself; contradicted by other information on the subject; confirmed by other independent sources.
7	Deception	Deliberately false; contradicted by other information on the subject; confirmed by other independent sources.
8	Cannot Be Judged	No basis exists for evaluating the validity of the information.

INTERNET SEARCH TECHNIQUES

A-20. Certain basic techniques and procedures can be applied to all Internet searches. Table A-5, page A-9, lists steps, techniques, and procedures for searching the Internet. *Untangling the Web: An Introduction to Internet Research* and similar works provide detailed discussion of the Internet and advanced Internet search techniques. Updated annually, *Untangling the Web* is a DOD product available online at http://www.osis.gov.

Table A-5. Internet search techniques and procedures

Step	Technique and Procedure
Plan Research	Use mission and specific information requirements to determine objective and search terms.
	Write all search terms down.
	Collaborate with other analysts to determine potential Internet sites.
	Select the search tools and sources that will best satisfy the objective.
	Comply with legal restrictions.
	Determine operations and computer security measures.
Conduct Search	Search by keywords.
	Search in natural language.
Refine Search	Compare the relevancy of the results to objective and indicators.
	Compare the accuracy of the results to search parameters (keywords, phrase, date or date range, language, format, and so on).
	Compare the results from different search engines to identify missing or incomplete information (for example, one engine's results include news articles but another engine's results do not).
	Modify the keywords.
	Search within results.
	Search by field.
	Search cached and archived pages.
	Truncate URLs.
Record Results	Bookmark or document Web page URL.
	Save content.
	Download files.
	Record citation.
	Identify intellectual property.
Assess Results	Evaluate source reliability.
	Evaluate information accuracy.

PLAN SEARCH

A-21. Intelligence personnel use their understanding of the supported unit's mission, specific information requirements, indicators, and the Internet to plan, prepare, and execute their search. The specific information requirement helps to determine what information to search for and where to look. The specific information requirement provides the focus and initial keywords to be used to search for information. Once identified, the analyst or collector records these search terms and analyzes potential domains, databases, organizations, or institutions with Web pages likely to yield results. Sites likely to yield unreliable or deceptive results can be excluded during planning. Once the parameters of the search and any known search target sites are established, a figure for man-hours expendable should be established, and personnel should be assigned specific areas to minimize duplication of effort.

SEARCH ENGINES

A-22. Search engines allow the user to search for text and images in millions of Web pages. The different commercial and government search engines vary in what they search, how they search, and how they display results. Most search engines use programs called Webcrawlers or spiders to build indexed databases. A Webcrawler searches Internet sites and files and saves the results in a database. The search engine, therefore, is actually searching an indexed database and not the content of the site or an online database. The search results also vary between search engines because each engine uses different Webcrawlers and searches different sites. Most engines display search results in order of a relevancy formula with a brief description and a hyperlink to the referenced Internet file or site.

Appendix A

A-23. Search engines use various search models and criteria called relevancy formulas. The relevancy formula evaluates how well the query results match the request. For Web pages that are commercially oriented, designing the page to achieve the highest ranking has become an art form. Some search sites practice "pay for placement." These sites simply sell the top slots in a search to the highest bidder. Other sites provide clearly delineated paid placement results from "pure" search results. Search engines are continually changing their relevancy formulas in order to try to stay ahead of Web developers. Some Web designers, however, load their sites with words like "free," "money," or "sex" in an attempt to influence the search engine's relevancy formula.

A-24. Other Web designers engage in practices called "spamdexing" or "spoofing" in an attempt to trick the search engine. The significance of the relevancy formulas to the user is the importance of understanding that the keyword in the search does not necessarily yield the same results with every search engine. This becomes obvious when the user considers that relevancy formulas vary from search engine to search engine and are in a constant state of evolution. In some formulas, the placement of the keywords yields different results if rearranged because the search engine's relevancy formula places more emphasis on the first words in the search string. Relevancy formulas may also assume importance depending on the type of search being done. For instance, a field search—which is limited to the Web page itself, such as title, URL, and date—may be more critical than a full-text search.

A-25. As search engines evolve, some engines have become adept at finding specific types of information such as statistical, financial, and news more effectively than other engines. To overcome this specialization, software engineers developed the metasearch engine. The metasearch engine allows the user to query more than one search engine at a time. On the surface this would seem to be the final answer to the search question; just query all search engines at one time. Unfortunately, it is not quite that easy. Since it must be designed to work with all search engines that it queries, the metasearch engine must strip out each search parameter to the lowest common denominator of each search engine. For example, if a particular search engine cannot accommodate phrases in quotation marks or a type of Boolean function, then the metasearch engine will eliminate that function from the search. The resulting search, in many instances, then becomes too broad and less useful than a well-formatted search using a search engine that the user is familiar with and that is known to be good at locating the type of information required. Regardless of the search engines used, there are four fundamental maxims applicable to all searches:

- Start with a specific term and expand to a general term only if necessary. For instance, if searching for an individual, begin with their name or most common alias, such as "Abu Sabaya," then progress as necessary to less-specific associations, such as "Abu Sayyaf" and then "Philippines terrorists."
- Use language and terms consistent with the probable sites that will produce "hits." In general, this means either using language of sufficiently high vocabulary and grammar or, as appropriate, using lower vocabulary, colloquialisms, or even deliberate grammatical errors.
- Be as specific as possible. For example, if searching for instances of a specific weapon, be as specific with nomenclature as possible; for example, use "RPG-7" and "Rocket-Propelled Grenade" instead of simply "RPG."
- Begin with the most-restrictive Boolean operators and progress as necessary to less-restrictive operators. For instance, if searching for proliferation of North Korean rocket technology to countries other than Iran, begin with a restrictive search, such as "IRFNA" and "North Korea" and "Taep'o-dong-2" not "Shahab."

Keyword Search

A-26. In keyword-based searches, the intelligence personnel should consider what keywords are unique to the information being sought. The analyst or collector needs to determine enough keywords to yield relevant results but not so many as to be overwhelmed with a mixture of relevant and irrelevant information. He should also avoid common words such as "a," "an," "and," and "the" unless these words are part of the title of a book or article. Most search engines ignore common words. For example, if looking for information about Russian and Chinese tank sales to Iraq, the analyst or collector should not use "tank" as the only keyword in the search. Instead, he should use additional defining words such as "Russian Chinese tank sales Iraq."

Open-Source Intelligence and Information

A-27. In most search engines, Boolean and math logic operators help the analyst or collector establish relationships between keywords that improve the search. However, some search engines default to certain Boolean operators and some may not allow the use of all operators. Using the operators listed in Table A-6, the search engine searches for Russian and tank together when the analyst or collector places the words within parentheses; for example, "(Russian tank)." If he wants to exclude "Chinese tank sales" from the search result, then he uses "(Russian tank) NOT (Chinese tank) sale Iraq" in the search. The analyst or collector can also use a NEAR search when the relationship and the distance between the terms are well established. For example, if the analyst or collector is looking for incidents of earthquakes in Pakistan, and news articles normally place the place name of the location of an attack within five words of "earthquake" in the title of the body of the article, then he uses "earthquake NEAR/5 Pakistan" in the search.

Table A-6. Boolean and math logic operators

Function	Boolean	Math	Example
Must be present*	AND	+	chemical AND weapon chemical + weapon
Must not be present	NOT	-	Africa NOT Sudan Africa - Sudan
May be present	OR	not applicable	chemical OR biological
Either of two terms but not both terms together on same document	XOR	#	anthrax XOR smallpox
Complete phrase	""	""	"Chinese tank sales to Iraq"
Nested	( )	not applicable	(Shining Path)
Near**	NEAR	not applicable	"White House" NEAR "airspace incursion"
Wildcards	word* or *word	not applicable	gun* (gunpowder, gun sight, and so on)
Stopwords***	"" ""	not applicable	""OR"" (do not ignore OR)

Notes:
*In some search engines, the default is AND. In this case, the analyst will have to use the OR operator or the equivalent option on a pull-down menu.
**Some engines use ten words as the distance between NEAR words. A forward slash and a number indicate the distance between the terms.
***Stopwords are words that search engines ignore because they are too common or are reserved for a special operation. There is no uniform list, but they include words such as an, any, to, with, and from. They also include the standard Boolean operators AND, NOT, NEAR, and OR.

NATURAL LANGUAGE SEARCH

A-28. An alternative to using a keyword search is the natural language question format. Most of the major search engines allow this capability. The analyst or collector obtains the best results when the question contains good keywords. One of the major downsides to this technique is the large number of results. If the needed information is not found in the first few pages, then the analyst should initiate a new search using different parameters.

REFINE THE SEARCH

A-29. Normally, the first few pages of search results are the most relevant. Based on these pages, the analyst or collector evaluates the initial and follow-on search to determine if the results satisfied the objective or if additional searches are required. During evaluation, he compares—

- Relevancy of the results to the objective and indicators.
- Accuracy of the results to search parameters (keywords, phrase, date or date range, language, format, and so on).
- Results from different search engines to identify missing or incomplete information (for example, one engine's results include news articles but another engine's results do not).

Appendix A

MODIFY THE KEYWORD

A-30. If initial search attempts are unsatisfactory, the analyst or collector can refine the search by changing—

- *Order*. Search engines may place a higher value or more weight on the first word or words in a multiple word or phrase search string. Changing the word order from "insurgents Iraq" to "Iraq insurgents" may yield different search results.
- *Spelling/Grammar*. Search engines attempt to match the exact spelling of the words in the search string. There are search engines that do recognize alternate spellings or prompt the user to correct common misspellings. Changing the spelling of a word from the American English "center" to the British English "centre" may yield different results. Changing the spelling of a transliterated name from "Al-Qaeda" to "al-Qaida," "al-Qa'ida," "el-Qaida," or "al Qaeda" generates different results that may be useful depending upon the objective of the search. Some search engines provide this capability for a "sounds like" type search that eliminates or reduces the manual entry of each variation. Looking for common misspellings or common grammatically incorrect short phrases may be useful in yielding results from a source for which English is a second language or the language of the Web page is in a second language for the Web designer or Web contributor.
- *Case*. Search engines may or may not support case-sensitive searches. Like spelling, some engines attempt to match the word exactly as entered in the search. The intelligence personnel should use all lowercase letters for most searches. When looking for a person's name, a geographical location, a title, or other normally capitalized words, then the intelligence personnel should use a case-sensitive search engine. Changing the case of a word from "java" to "Java" changes the search result from sites about coffee to sites about a programming language.
- *Variants*. Intelligence personnel use terms that are common to their language, culture, or geographic area. Using variants of the keyword, such as changing "policeman" to "cop," "bobby," "gendarme," "carabiniere," "policía," "polizei," or other form, may improve search results.

SEARCH WITHIN RESULTS

A-31. If the initial or follow-on search produces good but still unsatisfactory results, the analyst or collector can search within these results to drill down to the Web pages that have a higher probability of matching the search string and containing the desired information. Most of the popular search engines make this easy by displaying an option such as "search within these results" or "similar pages" that the user can select. Selecting the option takes the analyst or collector to Web pages with additional related information.

SEARCH BY FIELD

A-32. In a field search, the analyst or collector looks for the keywords within the URL as opposed to searching the entire Internet. The best time to use a field search is when the search engine returned a large number of Web pages. Although capabilities vary by search engine, some of the common field search operators are—

- *Anchor*: Searches for Web pages with a specified hyperlink.
- *Domain*: Searches for specific domains (Table A-7, page A-13, or the following URL—http://www.iana.org).
- *Like*: Searches for Web pages similar or related in some way to a specified URL.
- *Link*: Searches for a specific hyperlink embedded in a Web page.
- *Text*: Searches for specific text in the body of the Web page.
- *URL*: Searches for specific text in complete Web addresses.

Table A-7. Internet domains

Domain	Description	Operator/Sponsor
.aero	Reserved for members of the air-transport industry.	Société Internationale de Télécommunications Aéronautiques
.biz	Restricted to businesses.	NeuLevel, Inc.
.com	Unrestricted top-level domain intended for commercial content.	VeriSign Global Registry Services
.coop	Reserved for cooperative associations.	Dot Cooperation LLC
.edu	Reserved for postsecondary institutions accredited by an agency on the U.S. Department of Education's list of Nationally Recognized Accrediting Agencies.	Educause
.gov	Reserved exclusively for the U.S. Government.	U.S. General Services Administration
.info	Unrestricted top-level domain.	Afilias Limited
.int	Used only for registering organizations established by international treaties between governments.	Internet Assigned Number Authority
.jobs	Reserved for human resource managers.	Dot Cooperation LLC
.mil	Reserved exclusively for the U.S. military.	U.S. DOD Network Information Center
.museum	Reserved for museums.	Museum Domain Management Association
.name	Reserved for individuals.	Global Name Registry
.net	Intended for Internet Service Providers.	VeriSign Global Registry Services
.org	Intended for noncommercial use but open to all communities.	Public Interest Registry
.pro	Restricted to credentialed professionals and related entities.	RegistryPro

Note: Country code domain names are available at http://www.iana.org.
Source: Internet Assigned Number Authority at http://www.iana.org.

SEARCH BY DOMAIN

A-33. With the millions of URLs on the Web, the analyst or collector is faced with a myriad of sites that may or may not actually be produced and maintained by the type of organization represented by the majority of Web pages in that domain (Table A-7). Certain domains, such as ".mil," ".edu," and ".gov," are consistently reliable as being administered and authored by military, educational, or governmental organizations. Although not universally free of errors, the content of these sites can be consistently relied upon as vetted and factual. Although ".edu" sites contain content that is generally vetted by peer groups, certain forums on academic Web sites may contain content that is opinion, commentary, or satire. Researchers can generally infer whether content is presented as researched fact or as opinion.

A-34. Several domains have, over years of ever-increasing numbers of Internet participants, become highly suspect as to the validity of the organization using such a domain extension. In particular, the open-source information gatherer must not take ".org," ".info" or ".net" extensions as necessarily produced by a bona fide organization for that domain. Increasingly on these domains are sites whose agenda is profit. In addition, the .org domain is populated by organizations with actual members and individuals purporting to represent groups that essentially do not exist. Similarly, the lines of sites having a .com address are increasingly blurred. When looking at domains with large numbers of questionable sites, greater care in determining source reliability is required. This can be mitigated some by using recognized "brand name" sites; for instance, all major news sites have an editorial stance but the basic facts of a story from a major news organization or periodical can generally be assumed to be accurate. Similarly, most corporate Web sites from the major industrialized nations can be considered consistently reliable in terms of technical data and statistics. Major international governmental organizations and nongovernmental organizations, such as the International Committee of the Red Cross or the American Psychiatric Association, can generally be considered reliable.

Appendix A

SEARCH IN CACHE AND ARCHIVE

A-35. Sometimes a search or an attempt to search within results returns a URL that matches exactly the search objective but when the analyst or collector tries to link to the site, the link or the site is no longer active. If the search engine captures data as well as the URL, they can select a "cached" link to access the original data. Another technique is to search in an Internet archive site such as "www.archive.org" for the content. The analyst or collector needs to be aware that this information is historical and not subject to update by the original creators.

TRUNCATE THE UNIFORM RESOURCE LOCATOR

A-36. In addition to using the search engine to search within results, the analyst or collector can also manually search within the results by truncating the URL to a Web page. The analyst or collector works backward from the original search result to the Web page or home page containing the desired information or database by deleting the end segments of the URL at the "/" forward slash. This technique requires a basic understanding of how Web page designers structure their Web sites.

RECORD RESULTS

A-37. Intelligence personnel and other Army special operations Soldiers must save the search results that satisfy the research objective. In addition, a record/running log of unreliable and/or deceptive sites may be maintained. Saving the results enables the analyst or collector to locate the information later as well as properly cite the source of the information in intelligence reports and databases. Although printing a hard copy is an option, a soft-copy (electronic) record of the search results provides a more portable and versatile record. Some basic techniques for saving an electronic record of the search results are—

- *Bookmark or record source.* Bookmark the link to the Web page using the "bookmarks" or "favorites" option on the Internet browser and/or record the author or organization, title, publication or posting date, retrieval date, and URL of the information.
- *Save content.* Save all or a portion of the Web page content by copying and pasting the information in a text document or other electronic format such as a field within a database form. The naming convention for the soft-copy record should be consistent with unit electronic file management standards. As a minimum, the record should include the URL and retrieval date within the file.
- *Download files.* Download audio, image, text, video, and other files to the workstation. The naming convention for the soft-copy record should be consistent with unit electronic file management standards.
- *Save Web page.* Save the Web page by using the Internet browser's "save as" option and the ".mht" Web archive file type. Doing so creates a complete, stable record of the entire Web page. It may be necessary to include the date and time in the file name to ensure a complete citation for the information.
- *Identify intellectual property.* Identify intellectual property that an author or an organization has copyrighted, licensed, patented, trademarked, or otherwise preserved the rights to. Some Web pages list the points of contact and terms-of-use information at the bottom of the site's home page.

Appendix B
Target Analysis Outline

This appendix provides a sample target analysis outline (Figure B-1, pages B-1 through B-4). The outline serves as a site survey and is the baseline document for establishing the CARVER matrix. More information on CARVER evaluation criteria can be found in Appendix F of JP 3-05.1 and FM 3-05.

(CLASSIFICATION)

1. ( ) Administrative Data.
    a. ( ) Name of facility.
    b. ( ) Location (address).
        (1) ( ) Map coordinates (universal transverse mercator [UTM], latitude, and longitude).
        (2) ( ) Geographic area.
            (a) ( ) Urban.
            (b) ( ) Suburban.
            (c) ( ) Rural.
    c. ( ) Date of analysis.
    d. ( ) Author and sourcing.
    e. ( ) List of attachments.
        (1) ( ) Maps.
        (2) ( ) Photographs.
        (3) ( ) Brochures.
        (4) ( ) Schedules.
        (5) ( ) Sketches.
        (6) ( ) Blueprints.
2. ( ) General.
    a. ( ) General description of components of the facility and a brief comment on the nature and the importance of the operation.
    b. ( ) Description of the component parts and critical nodes of the facility.
        (1) ( ) Physical structure, sketch or photo (air or ground) with dimensions.
        (2) ( ) Communications.
            (a) ( ) Type.
            (b) ( ) Backup systems.
            (c) ( ) Command and control center.
        (3) ( ) Power and fuel.
            (a) ( ) Types used, primary and secondary.
            (b) ( ) Amount used.
                1. ( ) Daily rate.

(CLASSIFICATION)

Figure B-1. Sample target analysis outline

(CLASSIFICATION)

        2. ( ) Seasonal variation.
   (c) ( ) Sources of supply.
        1. ( ) On-site storage.
        2. ( ) Means of delivery and time required for resupply.
   (d) ( ) Type of storage facility.
        1. ( ) Aboveground.
        2. ( ) Underground.
        3. ( ) Combination.
   (e) ( ) Amount and type of fuel on hand.
   (f) ( ) Reserve system and conversion time.
        1. ( ) Type used.
        2. ( ) Amount used (consumption rate).
        3. ( ) Sources of supply.
           a. ( ) On-site storage.
           b. ( ) Means of delivery and time required to resupply.
        4. ( ) Type of storage facility.
           a. ( ) Aboveground.
           b. ( ) Underground.
           c. ( ) Combination.
(4) ( ) Personnel.
   (a) ( ) Number of employees.
   (b) ( ) Number of employees present during each shift.
   (c) ( ) Work hours or work days.
   (d) ( ) Key personnel.
   (e) ( ) Labor organizations and labor-management relations.
(5) ( ) Raw materials.
   (a) ( ) Type.
   (b) ( ) Amount.
        1. ( ) Daily, weekly, or monthly.
        2. ( ) Stockpiles.
   (c) ( ) Sources of supply.
   (d) ( ) Means of delivery.
(6) ( ) Finished product.
   (a) ( ) Type (flammable, nonflammable).
   (b) ( ) Amount (daily, weekly, or monthly production).
   (c) ( ) Quality control.
   (d) ( ) By-products.
        1. ( ) Types.
        2. ( ) Amount.
        3. ( ) Backup system.
        4. ( ) Maintenance.

(CLASSIFICATION)

Figure B-1. Sample target analysis outline (continued)

(CLASSIFICATION)
- (e) ( ) Distribution.
- (f) ( ) Stockpile storage and type of storage facility.
- (g) ( ) Conversion to manufacture of war materials.
- (7) ( ) Transportation and materials-handling equipment.
  - (a) ( ) Types.
  - (b) ( ) Amount.
  - (c) ( ) Backup systems.
  - (d) ( ) Maintenance and repair.
- (8) ( ) Flow diagrams.
- (9) ( ) Security.
  - (a) ( ) Types of systems.
    1. ( ) On-site.
    2. ( ) Reserve systems and reaction time.
  - (b) ( ) Type armament and circumstances under which used.
  - (c) ( ) Amount employed and scheduled.
  - (d) ( ) Location.
    1. ( ) On-site.
    2. ( ) Reserve system.
  - (e) ( ) Screening systems.
  - (f) ( ) Communications system.
  - (g) ( ) Crisis-control equipment and personnel.
    1. ( ) Type.
    2. ( ) Amount.
    3. ( ) Location.
    4. ( ) Reaction time.
    5. ( ) Emergency access.
    6. ( ) Alarm system.
    7. ( ) Medical facilities.
    8. ( ) Fire-fighting capabilities.
       a. ( ) Personnel.
       b. ( ) Equipment.
       c. ( ) Procedure.

3. ( ) Specifics.
   a. ( ) List all critical components within the complex.
   b. ( ) List all potential targets within the complex that meet current statement of requirements.
   c. ( ) List common targets within the complex.
   d. ( ) Describe the relationship of the target to related facilities or systems.
      (1) ( ) Internal.
      (2) ( ) External.

(CLASSIFICATION)

Figure B-1. Sample target analysis outline (continued)

(UNCLASSIFIED)

4. ( ) Conclusions.
   a. ( ) Based upon analysis of target complex, identify and justify the components deemed most susceptible to attack by—
      (1) ( ) A small force (1 to 12 men) with conventional weapons and munitions.
      (2) ( ) A large force (50+ men) with conventional weapons and munitions.
      (3) ( ) A small force with unconventional weapons and munitions.
      (4) ( ) An operator-in-place (underground).
   b. ( ) Determine consequent downtime and destructive effect such an attack would have against the target facility.

(UNCLASSIFIED)

Figure B-1. Sample target analysis outline (continued)

Appendix C
Interagency and Multinational Intelligence

Improving synergy of interagency intelligence in order to meet the needs of agencies collecting and using intelligence as they execute global operations against terrorism networks is a high priority. In every geographic area of responsibility close interagency cooperation is required. Several areas have emerged as consistent areas of interagency intelligence concern, including the timely sharing of CI, law enforcement information, and other actionable intelligence regarding asymmetric threats from terrorism, weapons of mass destruction, and information activities. In addition, few operations conducted against terrorism networks have been unilateral. Global operations against terrorism networks are inherently multinational as well as interagency in nature. Cooperation between both traditional and first-time multinational partners will continue to present challenges to Army special operations intelligence staffs. This appendix focuses on habitual and ad hoc relationships between interagency and multinational intelligence organizations and staffs, as well as the doctrine addressing interagency and multinational intelligence.

INTERAGENCY INTELLIGENCE

C-1. Interagency coordination is described in JP 3-08, *Interorganizational Coordination During Joint Operations*, as "more art than science." This is in contrast to military operations that depend on standing operating procedures and doctrine to provide structure. Increasingly, more and more personnel in other government agencies have operated in an interagency environment that has included DOD personnel. For some Army special operations units, coordination with other government agency intelligence personnel is well established and habitual. Examples of interagency and multinational organizations at the strategic, operational, and tactical levels are shown in Table C-1, page C-2.

C-2. The JIC is the primary focal point for providing intelligence support to a combatant command. The JIC must analyze theater intelligence production requirements, collection requirements, and requests for information from subordinate commands to determine whether such intelligence needs can be met with organic resources or may require national-level assistance. If the JIC determines national-level production assistance is required, a formal request will be prepared in the form of a request for information. The combatant command's staff may coordinate for the deployment of a NIST to help ensure JTF connectivity with the theater JIC and national intelligence agencies. Depending on the level of a SOTF in the command structure of an operation, the SOTF may request deployment of a NIST or coordinate national-level intelligence requests through the JTF. In addition, subordinate units may have liaison personnel from national-level intelligence and law enforcement organizations collocated with them.

C-3. The interagency support provided by a NIST allows access to agency-unique information and analysis. It affords a link to national-level databases and information that can provide information beyond the organic resources of the JTF. NIST members are available to the JTF and combatant command HQ before deployment for integration and mobilization training. Participating agencies retain control of their members deployed with the NIST, but the NIST operates under the supervision of the JTF intelligence staff.

C-4. Two distinct methods of requesting interagency intelligence support have been established in joint doctrine (JP 2-01). The noncrisis and crisis request procedures are discussed in the following paragraphs.

Appendix C

Table C-1. Interagency and multinational organizations at the strategic, operational, and tactical levels

	Executive Departments and Agencies	Armed Forces of the United States	State and Local Government	NATO	United Nations	Nongovernmental Organizations
Strategic	Department Secretaries Agency Directors CIA FBI Homeland Security	Secretaries of Defense Chairman, Joint Chiefs of Staff GCC	Governor Adjutant General	NATO HQ Supreme Allied Commander, Europe	UN HQ Functional HQ	International or National HQ President or Chief Executive Council
Operational	Ambassador/Embassy Staff Other Government Agency Liaisons Federal Coordinating Officer	JTF JSOTF Defense Coordinating Officer	State Coordinating Officer Emergency Management Department/Agency	Major Subordinate Commands	Special Representative to the Secretary General UN Command Korea (when activated)	Regional or Field Officers
Tactical	Ambassador/Embassy Staff Consul and Staff Other Government Agency Teams	Service Components	National Guard Units Local Officials Local Services	Principal Subordinate Commands Combined Joint Task Force Units	Special Representative to the Secretary General Force Commander Units/Observers	Field Teams Relief Workers

NONCRISIS REQUEST

C-5. Noncrisis request procedures include the following:
- The DIA ensures the flow of military intelligence from the national level through the JICs to deployed forces during peacetime.
- Requests for information are forwarded from the geographic combatant command joint intelligence operations center to the Defense Joint Intelligence Operations Center, Directorate for Intelligence, operational intelligence coordination center, and/or production agency.
- The JIC determines national-level intelligence collection requirements.
- Collection request is prepared and forwarded to DIA.

CRISIS REQUEST

C-6. Crisis request procedures include the following:
- The NMJIC receives crisis requests for information.
- The JIC forwards time-sensitive collection requirements to Defense Intelligence Center defense collection coordination center.
- The NIST serves as a direct link to the NMJIC request for information desk and defense collection coordination center on order of the joint force intelligence officer.
- The JIC or equivalent receives a simultaneous copy of all requests for information forwarded by the NIST.

C-7. The JISE is the primary intelligence apparatus of the JTF commander. It is established along with the JTF itself. The JISE may constitute a new entity, or it may be little more than the combatant command's JIC, or elements thereof, moving forward. The JISE will face unique challenges in providing adequate and appropriate support to the JTF. The element will have to meld traditional sources of classified military information with unclassified information from open sources and local HUMINT. The sensitivities of nonmilitary partners in interagency activities to the concept of military intelligence will complicate the melding process.

C-8. Consideration must be given to control of sensitive or classified military information in forums such as the CMO center that include representatives of indigenous populations and institutions, other government organizations, international government organizations, nongovernmental organizations, and regional organizations. Procedures for the control and disclosure of classified information practiced by the DOD often either do not exist within other agencies or may be significantly different from DOD procedures. This omission or difference may result in the inadvertent or intentional passage of sensitive information to individuals not cleared for access to such information.

C-9. The GCC has the authority and responsibility to control the disclosure and release of classified military information within the joint operational area or joint special operations area IAW Chairman of the Joint Chiefs of Staff Instruction (CJCSI) 5221.01C, *Delegation of Authority to Commanders of Combatant Commands to Disclose Classified Military Information to Foreign Governments and International Organizations*. In the absence of sufficient guidance, command intelligence staffs should share only information that is mission essential, affects lower-level operations, and is perishable. When required, the appropriate operational echelon should receive authority to downgrade a classification or to sanitize information. All U.S. classified information must be marked releasable prior to its discharge to an external agency. Many external organizations can only provide a minimal level of security protection to classified information given them by the United States. Therefore, it should be assumed that they will either intentionally or unintentionally disclose the information they receive to unauthorized individuals. The final determination on releasing classified information should therefore be based on this assumption.

C-10. When conducting homeland security operations, Army special operations units may submit requests for information directly or through a higher echelon to a Joint Interagency Intelligence Support Element. The Joint Interagency Intelligence Support Element is an interagency intelligence component within an FBI's joint operations center designed to fuse intelligence information from the various agencies participating in response to a weapon of mass destruction, terrorist, or other threat or incident. The Joint Interagency Intelligence Support Element is an expanded version of the investigative and intelligence component, which is part of the standardized FBI command post structure. The Joint Interagency Intelligence Support Element manages five functions: security, collections management, current intelligence, exploitation, and dissemination.

MULTINATIONAL OPERATIONS DOCTRINE

C-11. Doctrine for multinational operations is included in Allied joint publications. These manuals provide principles and concepts to operate in a multinational operation. The intelligence architecture discussed in Chapter 3 may provide a framework to build the multinational intelligence architecture. However, the architecture and systems of the U.S. intelligence community must be interfaced and accessed by multinational partners with caution and care as to those systems and that information that will be accessed and released to foreign nations and agencies. Commonality of terms and definitions in organizations comprised of exclusively NATO member nations or organizations with a preponderance of NATO signatories may be achieved through adoption of terms from the appropriate allied administrative publication. Multinational partnerships made up of non-NATO members must develop their own unique TTP. In these cases, U.S. or NATO doctrine may serve as a basic guide for these TTP.

JOINT AND MULTINATIONAL DOCTRINE RELATIONSHIP

C-12. There are close analogies between joint and multinational doctrine that stem from similar needs—to present a seamless force to an adversary and for unity of effort. Many of the principles, issues, and answers to joint operations will be the same or similar in multinational operations. For multinational doctrine, a need exists to understand differences in cultural and national perspectives to adapt doctrine or develop new doctrine.

MULTINATIONAL INTELLIGENCE PRINCIPLES

C-13. The principles discussed in the following paragraphs are considerations for building intelligence doctrine for multinational operations. These considerations are in addition to the appropriate principles found in joint intelligence doctrine.

ADJUST FOR DIFFERENCES AMONG NATIONS

C-14. A key to effective multinational intelligence is a readiness, beginning with the highest levels of command, to make required adjustments to national concepts for intelligence support and make the multinational action effective. Areas that need to be addressed include designating a single director of intelligence, adjusting those intelligence support differences that may affect the integrated employment of intelligence resources, and sharing intelligence and information. With these things done, successful intelligence support rests in the vision, leadership, skill, and judgment of the multinational command and staff groups.

MAXIMIZE UNITY OF EFFORT

C-15. Intelligence officers of each nation need to view the threat from multinational and national perspectives. When the multinational organization faces a common adversary, a threat to one international military partner by the common adversary should be considered a threat to all international military partners. Soldiers from other nations bring a variety of capabilities and procedures to a multinational intelligence staff. Synergizing both collection and analysis will ensure the minimizing of duplicate or divergent efforts. Intelligence efforts of the nations should be complementary. Because each nation will have intelligence system strengths and limitations or unique and valuable capabilities, the sum of intelligence resources and capabilities of the nations should be available for application to the whole of the intelligence problem.

FORGE SPECIAL ARRANGEMENTS

C-16. The multinational command and national forces' intelligence requirements, production, and use should be agreed on, planned, and exercised well in advance of operations. For anticipated situations and operations, a prime objective should be attaining compatibility of intelligence doctrine and concepts, intelligence systems, intelligence-related communications, language and terms, and intelligence services and products.

C-17. Solutions to problems should be developed and tried before their need in actual operations so that doctrine and procedures do not become a trial-and-error methodology during combat. The concepts and exercise programs of the NATO and the United States-Republic of Korea Combined Forces Command provide illustrations of multinational doctrine development and testing.

C-18. Special arrangements unique to international military partnerships should be considered for developing, communicating, and using intelligence where there are differences in nations' cultures, languages and terminology, organizations and structures, operating and intelligence concepts, methodologies, and equipment.

COORDINATE INTELLIGENCE SHARING

C-19. The nations should share all relevant and pertinent intelligence about the situation and adversary to attain the best-possible common understanding of threatened interests and to determine relevant and attainable objectives. If possible, the international partners should conceive and exercise the methodology for exchanging intelligence well before operations begin. They must, when necessary, monitor and adapt the exchange during operations to meet better-understood or changing circumstances. The Army special operations commander should have personnel knowledgeable in foreign disclosure policy and procedures, and should obtain necessary foreign disclosure authorization from the release authority as soon as possible. Assignment of personnel familiar with foreign disclosure regulations to the joint or multinational task force will facilitate the efficient flow of intelligence.

C-20. Sharing intelligence sources and methods, including cooperative intelligence collection and production, may help attain the common objectives of multinational partners. However, when intelligence sources and methods cannot be shared among nations, the intelligence should be provided after it is sanitized by effectively separating the information from the sources and methods used to obtain it. This sanitizing process must also be exercised in peacetime for both known and probable allies. Intelligence production agencies should consider use of tear lines to separate that intelligence and information within a given report that may be immediately disclosed to multinational partners.

OPERATE A MULTINATIONAL INTELLIGENCE CENTER

C-21. When there is a multinational command, a multinational intelligence center should be established. Creating the center provides a facility for the commander, the director of intelligence, and staffs to develop multinational intelligence requirements statements and acquire and fuse the nations' intelligence contributions. The multinational intelligence center should include a representative from all nations participating in the multinational operation.

MAXIMIZE LIAISON

C-22. During multinational operations, inherent problems exist due to differences in culture, languages, terms, doctrine, methodologies, and operational intelligence requirements. Maximizing intelligence liaison personnel among commands and among supporting and supported organizations minimizes these problems.

This page intentionally left blank.

Appendix D
Linguist Support

This appendix covers the selection of linguists and OPSEC concerns and measures involved when using external linguist support. Although simultaneous oral translation and translation of the written word are two distinct activities, this appendix uses the terms translator, interpreter, and linguist interchangeably. Clear distinction between the tasks of simultaneous oral translation and translation of the written word are made through context.

LINGUIST CATEGORIES

D-1. Linguists may come from various backgrounds and possess various types of experience. Ideally, the most qualified translator is a highly educated U.S. citizen, capable of translating from English to the target language and vice versa without losing meaning. Beyond technical ability, the ideal candidate also possesses a security clearance. Because of the high demand for linguists, such persons are typically not present in sufficient numbers. Therefore, categories are assigned to translators to sort out various levels of reliability and experience. The U.S. Government has established the Interagency Language Round Table oral proficiency interview scale to provide a structured and objective measure of a person's ability to understand and convey ideas in a target language. The following categories outline the types of translators; however, each translator as an individual may be better or worse than these generalizations:

- *Category I*: Have native proficiency in the target language (Levels 4 and 5) and an advanced working proficiency (Interagency Language Round Table Level 2+) in English. They may be locally hired or from a region outside the area of operations, including U.S. citizens. They do not require a security clearance. Typically, they are the least reliable when translating complex English. U.S. Army special operations units must frequently work with Category I linguists who do not possess a security clearance and must be compartmentalized away from both sensitive and classified information while completing translation duties that involve only unclassified materials.
- *Category II*: Are U.S. citizens screened by DOD personnel and are granted access to Secret-level material by the designated U.S. Government personnel security authority. They have native proficiency in the target language (Levels 4 and 5) and an advanced working proficiency (Interagency Language Round Table 2+) in English.
- *Category III*: Are U.S. citizens screened by DOD personnel and are granted either a Top Secret/SCI clearance or an interim Top Secret/SCI clearance by the designated U.S. Government personnel security authority. They meet a minimum requirement of Interagency Language Round Table Level 3. They are capable of understanding the essentials of all speech in a standard dialect. They must be able to follow accurately the essentials of conversation, make and answer phone calls, understand radio broadcasts and news stories, and give oral reports (both of a technical and nontechnical nature).

COORDINATING LINGUIST SUPPORT

D-2. The primary COA is to identify language requirements and identify organic personnel with the required capability. When the primary COA is not viable, several alternative COAs are available, including—

- Other special operations units.
- Conventional/other Service units.

Appendix D

- Other government agency personnel.
- Expatriate Americans.
- Contractors.

D-3. Special operations units may fulfill linguist requirements by providing personnel for deployments to fulfill operational needs. The originating unit benefits from this as their Soldiers gain the opportunity to develop their language. If collocated, units should make all attempts to coordinate language-trained personnel between units so long as the mission of the unit losing that Soldier does not suffer.

D-4. When attempting to obtain language-qualified or simply language-familiar personnel from either Service or U.S. organizations, intelligence staffs should be aware that obtaining long-term use of language-qualified personnel is typically counterproductive to the loaning organization's overall mission. However, short-term or single-mission use of language-qualified personnel may be in the best interests of both organizations. In an interagency operation, other government agency linguist support from personnel from organizations such as the Department of State may be integral to mission success and may be a "go/no-go" criterion for U.S. Army special operations unit participation.

D-5. American citizens live in the overwhelming majority of countries in the world. Many go in pursuit of business or educational opportunities. Others go to conduct charitable or missionary work. Still others visit foreign countries and live on the fringes of those societies. Expatriates present both an opportunity and challenges for use as linguists. Americans living in a foreign country to conduct business or research often possess a high degree of language skill as well as sufficient education to deal with complex technical translations. The challenge will frequently exist that these persons either do not have the free time, the inclination, or the authorization of a parent company or university to act in even a limited linguistic role. An exception to this might be a defense company. In this instance, the personnel in question may also possess a security clearance.

D-6. Intelligence personnel should be aware that American expatriates working for charitable organizations in foreign countries will typically not be inclined to offer their linguistic abilities for anything other than a humanitarian mission. Some charitable organizations by charter will not work with anyone (for example, any government agency) that might compromise the appearance of neutrality. Still others will not work with military organizations on conscientious grounds. Intelligence staffs should be aware of local bias against anyone engaged in a religious mission or religious charity or a charity identified with a single country. In addition to these expatriates, other American expatriates may live on the fringes of the society in the foreign countries. Some are involved in illegal, quasi-legal, or illicit activities. Many of these may have social problems that disqualify them from acting as linguists. All potential linguists, including expatriate Americans, must be vetted by CI personnel.

D-7. Contracted linguist support falls into two basic categories. Linguists provided by companies with habitual relationships with DOD who routinely provide linguist support in one or more of the Interagency Language Round Table categories make up one class of contractor support. The other class of contract linguist support is those linguists obtained through U.S. Army special operations logistic systems.

VETTING AND CLEARANCES

D-8. Unit or task force CI sections must screen all linguists used in support of Army special operations units. For contractors and others possessing a valid security clearance, this process may consist primarily of verifying that clearance. However, CI personnel serve as a first line of evaluation for potential fiduciary, ethnic, religious, political, or other potential conflict of interest that might compromise the integrity of special operations units.

D-9. Exhaustive CI screening of personnel without security clearances is of paramount importance. All foreign nationals who work for and on American installations are a potential threat to the lives, property, and critical information of the U.S. Government. Perhaps no group poses an inherently higher risk than foreign nationals possessing linguistic abilities. In addition to their ability to garner intelligence through the speech or documents they are translating, persons possessing linguistic abilities often have superior education and technical knowledge to those workers speaking only their native language.

PROACTIVE APPROACHES

D-10. Several proactive approaches to using linguists can significantly contribute to OPSEC. Limiting the types of information the linguist translates is the most operationally practical. Clearly this is critical in the case of foreign nationals without a security clearance. Too frequent use of a single translator increases the likelihood of compromise of essential elements of friendly information. The following paragraphs will detail TTP most appropriate to contractual linguists without clearances but may be applied to greater or lesser degrees to cleared linguists (to include American civilians) as well. Some of the approaches to maximizing OPSEC include—

- Use of an interpreter pool.
- Compartmentalization.
- Double-blind translations.
- Orientation and training.
- Management.

Interpreter Pool

D-11. Use of an interpreter and linguist pool enhances OPSEC by decreasing the potential release and detection of essential elements of friendly information, vetting the reliability of translators through multiple supervisors, and potentially protecting the linguist from too frequent exposure in a given location. To maximize the benefit of an interpreter pool, individual interpreters should work on as varied a rotation in as many areas as possible with as many different teams and supervisors as possible. The disadvantage to be weighed against this approach is the possible difficulty in building rapport between the interpreter and the team. Advantages include the ability for several teams and supervisors to observe the linguist. Problems that consistently arise (such as arguments or difficulties in a single neighborhood or with a single group) can be documented and appropriate action taken. Another advantage with using a diverse pool is the ability to decrease the likelihood of the targeting of a specific interpreter to potential retaliation by varying the timing and frequency of his use.

Compartmentalization

D-12. Having an interpreter translate only a specific portion of a document, speech, or interview can help to avoid giving out sufficient operational details to compromise a mission. Dividing information into components may serve to compartmentalize it and prevent the piecing together of information by a single linguist. To maximize this effect, the sections or subsections of text or speech that a linguist is given to translate should be varied. This is not possible in all instances (for example, the simultaneous translation of a negotiation), but this technique should be used whenever possible.

Double-Blind Translation

D-13. The preferred method of translation is a double-blind process in which one translator or translation team translates from English to the target language, and another translator or team retranslates the translation back to English. Discrepancies between the two reveal a shortfall in linguistic ability or a deliberate mistranslation. Discerning the difference between these two reasons is the responsibility of both military intelligence and special operations Soldiers.

D-14. The double-blind process is most readily applied to the translation of documents but may also be applied to the spoken word. When using an interpreter to conduct a simultaneous translation, the double-blind method may be applied through either the overt or clandestine recording of the conversation being translated. This recording is then given in whole or more preferably in part (without the interpreter's original translation) to another interpreter. He then translates the conversation from the foreign language to English. The two translations are then compared for discrepancies. Few linguistic expressions are so literal that they do not leave some latitude for minor differences. Care should be taken to determine if a difference in translation is minor and understandable, indicative of a deeper incompetence, or deliberate injection of erroneous words, phrases, or content.

Appendix D

Orientation and Training

D-15. Initial orientation of interpreters to their duties, standards of conduct and competency expected from them, techniques to be used, security protocols, and work ethics requirements may greatly reduce both management and security issues. The orientation may include the following:
- Security procedures and confidentiality requirements.
- Specific duties and responsibilities.
- Acceptable and prohibited techniques.
- Acceptable and unacceptable job performance.
- Consequences of subversion or espionage.

D-16. The most important aspects of the initial training of an interpreter are stressing the need for truthfulness at all times and stressing supervisor and supervised relationships between the Soldier and the linguist. A contract linguist from the indigenous population may initially have trouble adapting to being a supervised employee as his former status and position may have been in the elite of his culture. Numerous security problems may arise if a linguist mistakenly believes he is in charge. It is essential during orientation that a linguist is made aware of his status as an employee.

Management

D-17. Good management of the linguists within the interpreter pool serves to further reduce potential security concerns. Foreign cultures frequently do not have the same work ethic as found in many Western nations. This is further complicated by the fact that few groups outside of the military have higher expectations on issues such as punctuality as the military. Initially conveying and then enforcing standards of conduct on issues such as punctuality, maintaining confidentiality, and even seemingly small details such as appearance, help to foster an atmosphere of professionalism. A professional and accountable linguist pool will invariably have fewer physical and OPSEC issues.

D-18. The Center for Army Lessons Learned (CALL) Handbook Number 04-07, *(FOUO) Interpreter Operations Handbook*, includes a complete discussion of the TTP on the other facets of using interpreters.

Appendix E
Document Exploitation and Handling

Document exploitation is a vital information source in the development of the all-source intelligence picture. Unless planned for and carefully monitored, the volume of captured enemy documents in all operations can rapidly overwhelm a unit's capability to extract meaningful information.

DEFINITIONS

E-1. A document is any piece of recorded information, regardless of form. Documents include printed material such as books, newspapers, pamphlets, operation orders, and identity cards, as well as handwritten materials such as letters, diaries, and notes. Documents also include electronically recorded media such as computer files, tape recordings, and video recordings, and the electronic equipment that contains documents or other vitally important intelligence. Examples include hard drives, operating systems, and personal electronic devices, including phones, personal digital assistants, and global positioning system devices. A captured enemy document may be needed by several collection or exploitation activities at the same time, requiring copies to be made. Collectors must have ready access to copying equipment. Documents often must be evacuated through two different channels for proper exploitation, which also makes copying necessary. Such documents and equipment require special handling to assure that they are returned to their owners.

E-2. Document exploitation is the systematic extraction of information from threat documents for the purpose of producing intelligence or answering information requirements. A threat document has been in the possession of the threat, written by the threat, or is directly related to a future threat situation. Document exploitation can occur in conjunction with HUMINT collection activities or as a separate activity.

E-3. A captured enemy document is any document that was in the possession of an enemy force that subsequently comes into the hands of a friendly force, regardless of the origin of that document. There are three types of captured enemy documents:

- *Official*: Documents of government or military origin.
- *Identity*: Personal items, such as identification cards or books, passports, and driver licenses.
- *Personal*: Documents of a private nature, such as diaries, letters, and photographs.

E-4. Open-source documents are documents that are available to the general public, including, but not limited to, newspapers, books, videotapes, public records, and documents available on the Internet or other publicly available electronic media.

E-5. Source-associated documents are documents that are encountered on or in immediate association with a human source. These may include both official and personal documents. Documents associated with human sources are normally exploited, at least initially, during the interrogation or debriefing of the source. Interrogators typically use these documents during planning and preparation for interrogation of the associated EPW. These personal documents and source identification documents are therefore evacuated in conjunction with the associated source and sent through prisoner, detainee, or refugee evacuation channels rather than through intelligence channels. If the duplication capability exists, collectors should copy personal documents that contain intelligence information and evacuate the copy through intelligence channels. The original personal document should be evacuated with the detainee but not on his person until the HUMINT collector has exploited it. Collectors evacuate official documents through intelligence channels after initial exploitation. If possible, the collector will copy official documents and evacuate the copy with, but not on, the source.

Appendix E

OPEN-SOURCE DOCUMENT OPERATIONS

E-6. Open-source document operations are the systematic extraction of information from publicly available documents in response to command information requirements. Open-source document operations can be separate operations or can be included as part of other ongoing operations. Open-source documents are significant in the planning of all operations, especially during the execution of stability operations and civil-support operations. As well as hard data, open-source information can provide valuable background information on the opinions, values, cultural nuances, and other sociopolitical aspects in the operational environment. In evaluating open-source documents, collectors and analysts must be careful to determine the origin of the document and the possibilities of inherent biases contained within the document.

CAPTURED DOCUMENT OPERATIONS

E-7. One of the significant characteristics of operations is the proliferation of recordkeeping and communications by digital methods. The rapid and accurate extraction of information from these documents contributes significantly to the commander's accurate visualization of his battlefield. Documents may be captured on or in immediate association with EPWs and detainees, may be found on or turned in by refugees, line crossers, displaced persons, or local civilians, or may be found in abandoned enemy positions or anywhere on the battlefield.

REQUIREMENTS FOR A DOCUMENT EXPLOITATION OPERATION

E-8. The number of personnel required to conduct document exploitation varies with the echelon and with the volume of documents. Regardless of the size of the operation, certain basic functions must occur:

- Supervision and administration.
- Accountability.
- Screening.
- Security requirements.
- Translation.
- Exploitation and reporting.
- Quality control.

DOCUMENT EVACUATION AND HANDLING

E-9. The rapid evacuation and exploitation of documents is a shared responsibility. It originates with the capturing unit and continues to the complete extraction of pertinent information and the arrival of the document at a permanent repository, normally at the joint level, either within the theater of operations or outside of it. Documents captured in association with detainees and EPWs, with the exception of identity documents, are removed from the individual to ensure that documents of intelligence interest are not destroyed. These documents are evacuated through EPW evacuation channels with, but not on the person of, the detainee. With the exception of official documents, all documents are eventually returned to the detainee.

E-10. Captured enemy documents not associated with a detainee are evacuated through military intelligence channels, generally starting with the capturing unit's intelligence staff. Depending on the type of documents, they may eventually be evacuated to the National Center for Document Exploitation. HUMINT collectors and translators can extract information of intelligence interest from captured enemy documents at every echelon; they will make an attempt to exploit captured enemy documents within their expertise and technical support constraints. Collectors evacuate captured enemy documents to different elements based upon the information contained and the type of document concerned. For example, documents related to criminal activity may be evacuated to the nearest criminal investigative unit. Direct evacuation to an element outside the chain of command takes place at the lowest practical echelon but is normally done by the first military intelligence unit in the chain of command. Document evacuation procedures are outlined in Annex B (Intelligence) of the unit's operation order and standing operating procedures.

Document Exploitation and Handling

Actions by the Capturing Unit

E-11. Document accountability begins at the time the document comes into U.S. possession. Original documents must not be marked, altered, or defaced in any way. The capturing unit attaches a DD Form 2745 (Enemy Prisoner of War [EPW] Capture Tag), Part C, to each document. Only in the case where a capturing unit does not have the time or the manpower to mark each document because of ongoing combat operations should the capturing unit fill out one capture tag for a group of documents. In this case, the capturing unit should place the documents in a weatherproof container (box or plastic bag). The capturing unit should fill out two copies of the DD Form 2745, placing one copy inside the container and attaching one to the outside of the container. If these forms are not available, the capturing unit records the required data on any piece of paper. At a minimum, the capturing unit should record the information as follows:

- Time the document was captured as a date-time group.
- Place document was captured, including an 8-digit coordinate, and description of the location. This should be as detailed as time allows. For example, if a terrorist safe house was searched, documents might be labeled based on which room of the house they were in, which filing cabinet, which desk, and so forth.
- Identity of the capturing unit.
- Identity of the source from whom the document was taken, if applicable.
- Summary of the circumstances under which the document was found.

E-12. Document evacuation procedures are listed in Annex B (Intelligence) to the operation order. If the capturing unit does not contain a supporting HUMINT collection team, it forwards any captured enemy documents found on the battlefield directly to its intelligence staff. The intelligence staff extracts priority intelligence requirement information as practicable, ensures that the documents are properly tagged, and ensures speedy evacuation to the next-higher echelon through intelligence channels. Normally, a capturing unit will use any available vehicle, and, in particular, empty returning supply vehicles, to evacuate documents. Documents captured on or in association with detainees, including EPWs, should be tagged and removed from the detainee. They are evacuated with detainees to a military police escort unit or an EPW holding facility.

E-13. When large numbers of documents are captured in a single location, it is often more expedient for the capturing unit to request a document exploitation team or HUMINT collection team from the supporting military intelligence unit be sent to the documents rather than attempting to evacuate all the documents. This reduces the burden on the capturing unit, facilitates the rapid extraction of information, and enables the priority evacuation of documents of importance to higher echelons. This method should only be used if the capturing unit can adequately secure the documents until the arrival of the document exploitation team and if the battlefield situation and military intelligence resources permit the dispatch of a team. The capturing unit should include in its request the following:

- The identification of the capturing unit.
- Its location and the location of the documents.
- The general description of the document site (such as an enemy brigade HQ).
- The approximate number and type of documents.
- The presence of captured computers or similar equipment.

E-14. The military intelligence unit dispatching the team should notify the requesting team as soon as possible to provide them an estimated time of arrival and to coordinate the arrival of the team. There is no set time for how long any particular echelon may keep a document for study. The primary aim of speedy evacuation to the rear for examination by qualified document exploitation elements remains. Each echelon is responsible to prevent recapture, loss, or destruction of the captured enemy documents.

Actions by the First HUMINT Collection or Document Exploitation Unit

E-15. The first HUMINT collection or document exploitation unit to receive captured enemy documents should log, categorize, and exploit the documents to the best of its abilities based on mission variables. They should rapidly identify documents requiring special handling or special expertise to exploit and evacuate those documents to the appropriate agencies. The military intelligence unit standing operating procedure or operation order should list special document evacuation requirements and priorities.

Appendix E

Accountability

E-16. The capturing unit and each higher echelon take steps to ensure that they maintain captured enemy document accountability during document evacuation. To establish accountability, the responsible element inventories all incoming captured enemy documents. Anyone who captures, evacuates, processes, or handles captured enemy documents must maintain accountability. All captured enemy documents should have completed captured document tags. An incoming batch of documents should include a captured document transmittal.

E-17. The exact format for a document transmittal is a matter of local standing operating procedure, but it should contain the information listed below:

- The identity of the element to which the captured enemy documents are to be evacuated.
- The identity of the unit forwarding the captured enemy documents.
- The identification number of the document transmittal.
- Whether or not captured enemy documents in the package have been screened and the screening category. (If not screened, NA is circled.) Document handlers should package documents that have been screened separately, by category.
- A list of the document serial numbers of the captured enemy documents in the package.

E-18. When a batch is received without a transmittal, the HUMINT collection element contacts the forwarding units and obtains a list of document serial numbers (if possible). The HUMINT collection element records all trace actions in its journal. Accountability includes—

- Inventorying the captured enemy documents as they arrive.
- Initiating necessary trace actions.
- Maintaining the captured document log (Figure E-1).

E-19. When a collector includes intelligence derived from a captured enemy document in an intelligence report, he references the identification letters and number of the document concerned to avoid false confirmation.

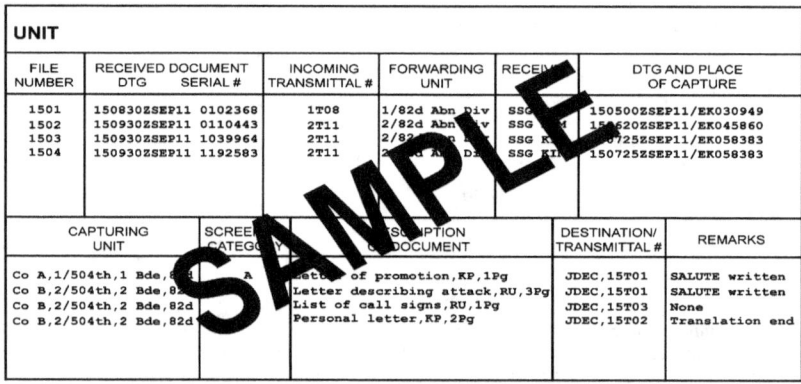

Figure E-1. Example of a captured document log

Inventory

E-20. The receiving element conducts an initial inventory of incoming captured enemy documents by comparing the captured enemy document to the captured document tag and accompanying transmittal documents. This comparison identifies—

- Transmittals that list missing captured enemy documents.
- Document tags not attached to captured enemy documents.

- Captured enemy documents not attached to document tags.
- Captured enemy documents not listed on the accompanying transmittal documents.

Trace Actions

E-21. The receiving unit initiates trace actions on all missing captured enemy documents, missing captured document tags, and all information missing from the captured document tags. They initiate trace actions by contacting elements from which the documents were received. The receiving unit can complete this corrective action swiftly if that unit's captured document log was filled out completely. If necessary, the trace action continues to other elements that have handled the document. If a captured document tag is unavailable from elements that have previously handled the captured enemy document, the document examiner fills out a captured document tag for the document using whatever information is available. Attempts to obtain missing captured enemy documents are critical because of the information those captured enemy documents might contain.

Document Logs

E-22. The captured document log is a record of what an element knows about a captured enemy document. After trace actions are initiated for any missing documents, the captured enemy documents are entered in the REMARKS section of the captured document log. This log must contain the following:

- Name of capturing unit.
- File number (a sequential number to identify the order of entry).
- Date-time group the captured enemy document was received at this element.
- Document serial number of the captured document tag.
- Identification number of the transmittal document accompanying the captured enemy document.
- Complete designation of the unit that forwarded the captured enemy document.
- Name and rank of individual that received the captured enemy document.
- Date-time group and place of capture (as listed on the captured document tag).
- Identity of the capturing units (as listed on the captured document tag).
- Document category (after screening).
- Description of the captured enemy document. (At a minimum, the description includes the original language; number of pages; type of document, such as a map, letter, or photograph; and the enemy's identification number for the captured enemy document, if available.)
- Destination and identification number of the outgoing transmittal.
- Remarks, to include any other information that can assist the unit in identifying the captured enemy document, including processing codes. These are set up by local standing operating procedures to denote all actions taken with the document while at the element, including intelligence reports, translations, reproductions, or return of the captured enemy document to the source from whom it was taken.

DOCUMENT SCREENING

E-23. Document screening is the rapid but systematic evaluation of documents to determine which documents contain priority information. Selected priority documents will be exploited immediately for priority intelligence requirement information and evacuated expeditiously (often electronically) to a document exploitation facility. Document screening can be done manually (requiring a linguist who is well-versed in the current collection requirements) or through the use of scanning devices with keyword identification capability. Document processing does not require the complete translation of a document but requires sufficient translation to determine the significance of the document.

E-24. As each document is screened, personnel assign one of four category designations to each document. The assigned category determines the document's priority for exploitation and evacuation. Document screening requires that the screening units receive the most current priority intelligence requirement and intelligence requirements, current friendly and enemy situation update, and relevant order of battle information. Screeners at higher echelons can recategorize captured enemy documents to accurately reflect the requirements at that level or information that has changed in category due to time sensitivity.

Appendix E

DOCUMENT CATEGORIES

E-25. Documents are divided into categories to prioritize their evacuation and the extraction of information from them for intelligence purposes. Document categories are discussed below.

Category A

E-26. Category A documents are those that require priority evacuation and/or special handling because of their special intelligence value. They contain size, activity, location, unit, time, and equipment (SALUTE)-reportable information. Category A documents also include those that are of interest to another command or agency (for example, TECHINT, Air Force, Navy, MISO units, or cryptographers).

E-27. The criteria for determining Category A documents changes according to the operational environment and will be set forth in each document exploitation element's standing operating procedure and in Annex B (Intelligence) of the unit's operation order. Documents that are evidence in legal proceedings against captured personnel suspected of crimes against humanity and war crimes will be handled as Category A documents. All Category A documents are handled as Secret. Standard Category A documents include, but are not limited to—

- Unmarked maps.
- Maps and charts containing any operational graphics, which are sent to the intelligence staff. From the intelligence staff, they would be evacuated to the all-source analysis center.
- Air Force-related documents, which should go to the nearest Air Force HQ.
- Navy-related documents, which should be sent to the nearest Navy HQ.
- TECHINT-related documents, which are evacuated to the nearest TECHINT unit.
- Cryptographic and communications-related documents, which are evacuated to the nearest SIGINT analysis unit.
- Documents constituting evidence to be used in legal proceedings against persons suspected of crimes against humanity and war crimes, which will be marked "CRIMINAL EVIDENCE." Such documents will be kept separate from other documents and will be stored under guard or in a secure area until turned over to a war crimes investigative unit. The SJA should be consulted concerning chain-of-custody requirements.

Category B

E-28. Category B documents contain information of intelligence interest to the supported command. The lowest echelon possible exploits the documents and evacuates them through intelligence channels. Category B documents are handled as Secret.

Category C

E-29. Category C documents and items contain no information of intelligence interest but still require special administrative accountability (for example, currency, works of art, narcotics). Currency is accounted for on DA Form 4137 (Evidence/Property Custody Document).

Category D

E-30. Category D documents contain no information of intelligence value. Only the theater or higher document repository can categorize documents as Category D.

GROUPED DOCUMENTS

E-31. Captured enemy documents are first grouped according to their assigned screening category. Personnel must be careful when sorting captured enemy documents to ensure no document is separated from its associated documents. These large groupings can be broken down into smaller groups. Each of these smaller groupings consists of captured enemy documents that were—

- Captured by the same unit.
- Captured in the same place.
- Captured on the same day at the same time.
- Received at the document exploitation element at the same time.

Document Exploitation and Handling

TRANSMITTAL OF CAPTURED ENEMY DOCUMENTS FROM FIRST AND SUBSEQUENT MILITARY INTELLIGENCE UNITS

E-32. Unless they have a HUMINT collection team in direct support, most units that capture or find documents normally have no way of evaluating, categorizing, or otherwise differentiating documents. They are all tagged and evacuated together by the most expedient means through military intelligence channels. Once these documents arrive at a HUMINT collection or document exploitation unit, the unit can screen, categorize, and extract information from the documents. The degree that documents are exploited at each echelon is dependent on mission priorities and available resources. Document handlers must note any attempts to exploit captured enemy documents on the transmittal documents to prevent unnecessary duplication of effort by higher echelons.

E-33. When transportation assets are limited, captured enemy documents are evacuated according to priority based on document categorization. All Category A captured enemy documents will be evacuated first, followed in order by Categories B, C, and D. Documents that have not yet been screened will be evacuated as Category C documents, but the transmittal slip will clearly indicate that the documents have not been screened.

E-34. Documents will be evacuated IAW unit standing operating procedure and Annex B (Intelligence) in the unit operation order. Lower-priority captured enemy documents, no matter how old, are never evacuated ahead of those with higher priority. Captured enemy documents are packaged so that a package of documents contains only those of one category. If the captured enemy document cannot be screened because of time or language constraints, it should be treated as a Category C, but kept separate from screened Category C captured enemy documents.

E-35. When captured enemy documents are evacuated from any echelon, a document transmittal sheet is used (Figure E-2, page E-5). A separate transmittal document is prepared for each group of captured enemy documents to be evacuated. The sending unit prepares a separate transmittal document for each separate addressee. The transmittal identification number is recorded in the captured document log (Figure E-3, page E-5) as part of the entry for each captured document. Copies of all translations should accompany the documents to avoid duplication of effort. If the sending unit submitted intelligence reports electronically, it should note the report number or include a copy of the report with the document to avoid duplicate reporting.

E-36. All captured enemy documents being evacuated must be accompanied with the appropriate—
- Technical document cover sheet.
- Secret cover sheet on Categories A and B documents.
- Translation reports and hard-copy reports accompanying translated documents.
- Captured document tags.

JOINT DOCUMENT EXPLOITATION FACILITY

E-37. The theater military intelligence brigade or group is normally tasked with the establishment of the theater joint document exploitation facility. The joint document exploitation facility is staffed by Army linguists, supported by technical experts from the Army and from the other Services, and supplemented as required by military and civilian contract translators. The joint document exploitation facility will normally contain military intelligence experts from SIGINT, CI, TECHINT, and other areas as required to identify and exploit documents of interest to these specialties.

E-38. Military and civilian translators must have security clearances appropriate to their mission requirements. This normally equates to at least a Secret clearance since the translators must be made aware of U.S. collection requirements to facilitate their work. The joint document exploitation facility performs a final examination of all documents of possible theater intelligence value before storing or evacuating them. The DIA sets procedures for exploitation of documents above theater-Army level.

DOCUMENT PROCESSING (RECOVERY AND TRANSLATION)

E-39. Units must normally process documents prior to exploiting them. Document processing includes the translation of foreign language documents into English, the recovery of damaged documents, the

Appendix E

decryption of encrypted documents, and the extraction of documents from electronic media. This need for processing frequently limits the amount of document exploitation that can be done outside a document exploitation facility.

DOCUMENT RECOVERY

E-40. At a minimum, the joint document exploitation facility manning includes teams trained in extracting and downloading information from electronic media such as computer hard drives. These individuals work in conjunction with TECHINT personnel responsible for the evaluation of captured computer hardware and software. These teams are prepared to deploy forward as necessary to accomplish their mission.

DOCUMENT TRANSLATION

E-41. Translations are not intelligence information reports. They are, however, often a precondition for document exploitation. Once translated, intelligence information can be extracted and reported. A translation should accompany the original foreign language document; a copy of the translation should accompany any copies of the original foreign language document and, as required, the intelligence reports. A translation report should contain the following information:

- Where the report will be sent.
- Which element prepared the report.
- Date-time group of the document translation.
- Report number as designated by local standing operating procedures.
- Document number taken from the captured document tag.
- Document description, including type of document, number of pages, physical construction of document, and enemy identification number, if applicable.
- Original captured document language.
- Date-time group document was received at element preparing the report.
- Date-time group document was captured.
- Place document was captured.
- Identity of capturing unit.
- Circumstances under which document was captured.
- Name of translator.
- Type of translation: full, extract, or summary.
- Remarks for clarification or explanation, including the identification of the portions of the document translated in an extract translation.
- Classification and downgrading instructions IAW AR 380-5, *Department of the Army Information Security Program*.

TYPES OF TRANSLATION

E-42. There are three types of translations:

- *Full*—one in which the entire document is translated. This is both time- and manpower-intensive, especially for lengthy or highly technical documents. Normally only a document exploitation facility at theater or national level will do full translations, and then only when the value of the information, technical complexity, or political sensitivity of the document requires a full translation. Even when dealing with Category A documents, it may not be necessary to translate the entire document to gain the relevant information it contains.
- *Extract*—one in which only a portion of the document is translated. For instance, a TECHINT analyst may decide that only a few paragraphs in the middle of a 600-page helicopter maintenance manual merit translation, and that a full translation is not necessary. The analyst would request only what he needed.
- *Summary*—one in which a translator begins by reading the entire document. He then summarizes the main points of information instead of rendering a full or extract translation. A

summary translation is normally written, but may be presented orally, particularly at the tactical level. Summary translations may be done as part of the document screening process. A summary translation requires a translator have more analytical abilities. The translator must balance the need for complete exploitation of the document against time available in combat operations. Translators working in languages of which they have a limited working knowledge may also use a summary translation. For instance, a Russian linguist may not be able to accurately deliver a full translation of a Bulgarian language document. However, he can probably render a usable summary of its content.

TRANSLATOR SUPPORT TO DOCUMENT EXPLOITATION

E-43. When HUMINT collectors are not available because of shortages or other mission requirements, document exploitation can be performed by military or civilian linguists under the management of a cadre of HUMINT collectors. Appendix D has further information on linguist support.

REPORTING

E-44. Information collected from documents is normally reported in a SALUTE report or an intelligence information report. Reporting through other reporting formats is discouraged. Intelligence reports are normally forwarded electronically or as otherwise directed by standing operating procedures and operational instructions. Normally, an electronic or hard-copy file of each report is maintained at the unit of origin, one electronic or hard copy is submitted through intelligence reporting channels, and one is forwarded with evacuated documents to the next unit to receive the document to prevent redundant reporting. In the event that the document itself cannot be evacuated in a timely manner, a verified copy of a translation report can be forwarded separately from the original document to an exploitation agency.

DOCUMENT EXPLOITATION

E-45. Documents found on detainees, including EPWs, that can be exploited more efficiently when combined with HUMINT collection are forwarded with the detainee to the next echelon in the EPW/detainee evacuation channel. In exceptional cases, documents may be evacuated ahead of the EPW or other detainee for advance study by intelligence units. A notation should be made on the EPW's capture tag or accompanying administrative papers about the existence of such documents and their location if they become separated from the detainee.

SOURCE-ASSOCIATED DOCUMENTS

E-46. Documents captured on or in association with a human source play an important role in the HUMINT collection process. These documents may contain reportable information the same as with any other captured enemy document. The information is immediately extracted from the documents and forwarded in the appropriate intelligence report. In addition to reportable information, documents may provide valuable insight into the attitude and motivation of the source and can be effectively used by the HUMINT collector in the approach process. Guidelines for the disposition of the detainee's documents and valuables are set by international agreements and discussed in more detail in AR 190-8, *Enemy Prisoners of War, Retained Personnel, Civilian Internees and Other Detainees*, and FM 3-19.4, *Military Police Leaders' Handbook*.

E-47. The capturing unit removes all documents, with the exception of the source's primary identification document, from an EPW or other detainee to prevent their destruction. These are placed in a waterproof container and Part C of the capture tag is placed in the bag. Documents from each source should be placed in a separate bag. These documents are turned over to the first military police EPW handling unit in the chain of command. The military police personnel will inventory all documents and prepare a hand receipt and provide a copy to the EPW or detainee.

E-48. To ensure proper handling and expeditious disposition of these documents, the first HUMINT collection element to see the detainee should review the documents as part of the source screening process. If an official document is confiscated and evacuated through military intelligence channels, the HUMINT collector must obtain a receipt for that document from the military police. If possible, the HUMINT

Appendix E

collection unit copies any documents that contain information of intelligence interest and forwards the copies through military intelligence channels. With the exception of an identification document, documents are normally kept separate from the detainee until the detainee arrives at a permanent confinement facility, at which time documents are returned to them per AR 190-8.

ACTIONS

E-49. Three possible actions may be taken with documents captured with a source. The documents may be confiscated, impounded, or returned to the source.

Confiscated

E-50. Documents confiscated from a source are taken away with no intention of returning them. Official documents, except identification documents, are confiscated and appropriately evacuated. The intelligence value of the document should be weighed against the document's support in the HUMINT collection of the source. The HUMINT collector must comply with the accounting procedures established for captured enemy documents by the military police personnel IAW AR 190-8.

Impounded

E-51. Some captured enemy documents will contain information that must be exploited at higher echelons. These documents may be impounded by the HUMINT collector and evacuated through intelligence channels. The HUMINT collector must issue a receipt to the source for any personal documents that he impounds. He must comply with the accounting procedures established for captured enemy documents impounded by the military police personnel IAW AR 190-8. When a captured enemy document is impounded, it is taken with the intent of eventual return. Personal documents with military information will be impounded if the military value is greater than the sentimental value. An example of a personal document whose military value might outweigh the sentimental value could be a personal photograph that includes military installations or equipment.

E-52. When a captured enemy document is impounded, it must be receipted. The receipt will include an itemized list of all the items taken from the prisoner and the name, rank, and unit of the person issuing the receipt. Items of high value may be impounded for security reasons. For instance, an EPW or detainee apprehended with an unusually large amount of money would have the money impounded and receipted. The military police will establish and maintain a DA Form 4237-R (Detainee Personnel Record) for impounded items. The register will identify the owner of the impounded items and provide a detailed description of the items impounded. A receipt will be given to anyone who has items impounded. AR 190-8 has procedures on handling personal effects.

Returned

E-53. Returned captured enemy documents are usually personal in nature. They are taken only to be inspected for information of interest and are given back to the source. Personal documents belonging to a source will be returned to the source after examination IAW the Geneva Convention Relative to the Treatment of Prisoners of War. These documents are captured enemy documents whose sentimental value outweigh their military value and may be returned to the source. Copies of these documents may be made and forwarded if necessary. Except for an identification document (which is always returned to the source), documents are evacuated with the source, but not physically attached to or on the source, until the source reaches a permanent confinement facility at echelons above corps.

Glossary

ADP	Army doctrine publication
ADRP	Army doctrine reference publication
AKO	Army Knowledge Online
AR	Army regulation
ARSOF	Army special operations forces
ASAS	All-Source Analysis System
ASAS-L	All-Source Analysis System-Light
ASAS-SS	All-Source Analysis System-Single Source
ASCOPE	areas, structures, capabilities, organizations, people, and events
ASK	Asymmetric Software Kit
ASPS	All-Source Production Section
ATP	Army techniques publication
BMATT	Briefcase Multimission Advanced Tactical Terminal
BRITE	Broadcast Request Imagery Technology Environment
C-2X	coalition CI/HUMINT staff element
CA	Civil Affairs
CALL	Center for Army Lessons Learned
CAO	Civil Affairs operations
CARVER	criticality, accessibility, recuperability, vulnerability, effect, and recognizability
CHATS	Counterintelligence/Human Intelligence Automated Tool Set
CHIMS	Counterintelligence/Human Intelligence Information Management System
CI	counterintelligence
CIA	Central Intelligence Agency
CMO	civil-military operations
COA	course of action
COLISEUM	Community On-Line Intelligence System for End-Users and Managers
CONUS	continental United States
DA	Department of Army
DCGS-A	Distributed Common Ground System-Army
DIA	Defense Intelligence Agency
DIA/DT	DIA Directorate for MASINT and Technical Collection
DISO	defense intelligence support office
DOD	Department of Defense
DODD	Department of Defense Directive
DTIC	Defense Technical Information Center
DTSS-D	Digital Topographic Support System-Deployable
EIW	Enhanced Imagery Workstation
ELINT	electronic intelligence

Glossary

ERDAS	Earth Resources Data Analysis System
EPW	enemy prisoner of war
F3EA	find, fix, finish, exploit, and analyze
FBI	Federal Bureau of Investigation
FM	field manual
FOUO	For Official Use Only
G-2	Assistant Chief of Staff, Intelligence
G-2X	CI/HUMINT staff element
G-3	Assistant Chief of Staff, Operations and Plans
G-7	Assistant Chief of Staff, Information Operations
G-8	Assistant Chief of Staff, Resource Management
G-9	Assistant Chief of Staff, Civil-Military Operations
GCC	geographic combatant commander
GEOINT	geospatial intelligence
GI&S	geospatial information and services
HN	host nation
HQ	headquarters
HQDA	Headquarters, Department of the Army
HUMINT	human intelligence
IAW	in accordance with
IMETS-L	Integrated Meteorological System-Light
IMINT	imagery intelligence
INSCOM	United States Army Intelligence and Security Command
INTELINK	intelligence link
IPB	intelligence preparation of the battlefield/battlespace
J-2	intelligence directorate of a joint staff
J-2O	Deputy Directorate for Crisis Operations
J-2X	CI/HUMINT joint staff element
JDISS	Joint Deployable Intelligence Support System
JDISS-SOCRATES	Joint Deployable Intelligence Support System-Special Operations Command, Research, Analysis, and Threat Evaluation System
JFC	joint force commander
JIC	joint intelligence center
JIPOE	joint intelligence preparation of the operational environment
JISE	joint intelligence support element
JP	joint publication
JSOTF	joint special operations task force
JTF	joint task force
JTT-B	Joint Tactical Terminal-Briefcase
JWICS	Joint Worldwide Intelligence Communications System
MASINT	measurement and signature intelligence

Glossary

MATT	Multimission Advanced Tactical Terminal
MDMP	military decisionmaking process
MID	military intelligence detachment
MIDB	Modernized Integrated Database
MIS	Military Information Support
MISG	Military Information Support group
MISG(A)	Military Information Support group (airborne)
MISO	Military Information Support Operations
MISOC	Military Information Support Operations Command
NATO	North Atlantic Treaty Organization
NGA	National Geospatial-Intelligence Agency
NGIC	National Ground Intelligence Center
NIST	National Intelligence Support Team
NMJIC	National Military Joint Intelligence Center
NSA	National Security Agency
OCONUS	outside the continental United States
OPSEC	operations security
OSINT	open-source intelligence
PMESII-PT	political, military, economic, social, information, infrastructure, physical environment, and time
PN	partner nation
POAS-MC	Psychological Operations Automated System – Message Center
PSYOP	Psychological Operations
S&TI	scientific and technical intelligence
S-2	intelligence staff officer
S-2X	CI/HUMINT staff officer
S-3	operations staff officer
SALUTE	size, activity, location, unit, time, and equipment
SCI	sensitive compartmented information
SCIF	sensitive compartmented information facility
SF	Special Forces
SFODA	Special Forces operational detachment A
SIGINT	signals intelligence
SIPRNET	SECRET Internet Protocol Router Network
SJA	Staff Judge Advocate
SOA	special operations aviation
SOAR	Special Operations Aviation Regiment
SOCRATES	Special Operations Command, Research, Analysis, and Threat Evaluation System
SODARS	Special Operations Debriefing and Retrieval System
SOF	special operations forces

Glossary

SOT-A	special operations team A
SOTF	special operations task force
SOWT	special operations weather team
SR	special reconnaissance
TACLAN	tactical local area network
TARP	Threat Awareness and Reporting Program
TCAE	Technical Control and Analysis Element
TECHINT	technical intelligence
TEMPEST	terminal electromagnetic pulse escape safeguard technique
TES-LITE	Tactical Exploitation System–Low-Cost Intelligence Tactical Equipment
TS-LITE	TROJAN Special Purpose Integrated Remote Intelligence Terminal Lightweight Integrated Telecommunications Equipment
TSOC	theater special operations command
TTP	tactics, techniques, and procedures
UAS	unmanned aircraft system
UN	United Nations
URL	Uniform Resource Locator
U.S.	United States
USAF	United States Air Force
USAJFKSWCS	United States Army John F. Kennedy Special Warfare Center and School
USAR	United States Army Reserve
USASFC(A)	United States Army Special Forces Command (Airborne)
USASOAC	United States Army Special Operations Aviation Command
USASOC	United States Army Special Operations Command
USATEC	United States Army Topographic Engineering Center
USSOCOM	United States Special Operations Command
UW	unconventional warfare

SECTION II– TERMS

all-source intelligence

    1. Intelligence products and/or organizations and activities that incorporate all sources of information, most frequently including human intelligence, imagery intelligence, measurement and signature intelligence, signals intelligence, and open-source data in the production of finished intelligence. 2. In intelligence collection, a phrase that indicates that in the satisfaction of intelligence requirements, all collection, processing, exploitation, and reporting systems and resources are identified for possible use and those most capable are tasked. (JP 2-0)

center of gravity

    The source of power that provides moral or physical strength, freedom of action, or will to act. (JP 5-0)

clandestine

    Any activity or operation sponsored or conducted by governmental departments or agencies with the intent to assure secrecy and concealment. (JP 2-01.2)

Glossary

counterintelligence

Information gathered and activities conducted to identify, deceive, exploit, disrupt, or protect against espionage, other intelligence activities, sabotage, or assassinations conducted for or on behalf of foreign powers, organizations or persons or their agents, or international terrorist organizations or activities. Also called **CI**. (JP 2-01.2)

critical capability

A means that is considered a crucial enabler for a center of gravity to function as such and is essential to the accomplishment of the specified or assumed objective(s). (JP 5-0)

critical requirement

An essential condition, resource, and means for a critical capability to be fully operational. (JP 5-0)

critical vulnerability

An aspect of a critical requirement which is deficient or vulnerable to direct or indirect attack that will create decisive or significant effects. (JP 5-0)

electronic intelligence

Technical and geolocation intelligence derived from foreign noncommunications electromagnetic radiations emanating from other than nuclear detonations or radioactive sources. Also called **ELINT**. (JP 3-13.1)

foreign instrumentation signals intelligence

A subcategory of signals intelligence, consisting of technical information and intelligence derived from the intercept of foreign electromagnetic emissions associated with the testing and operational deployment of non-US aerospace, surface, and subsurface systems. Foreign instrumentation signals include but are not limited to telemetry, beaconry, electronic interrogators, and video data links. (JP 2-01)

geospatial intelligence

The exploitation and analysis of imagery and geospatial information to describe, assess, and visually depict physical features and geographically referenced activities on the Earth. Geospatial intelligence consists of imagery, imagery intelligence, and geospatial information. Also called **GEOINT**. (JP 2-03)

human intelligence

A category of intelligence derived from information collected and provided by human sources. Also called **HUMINT**. (JP 2-0)

imagery intelligence

The technical, geographic, and intelligence information derived through the interpretation or analysis of imagery and collateral materials. Also called **IMINT**. (JP 2-03)

information environment

The aggregate of individuals, organizations, and systems that collect, process, disseminate, or act on information. (JP 3-13)

intelligence preparation of the battlefield/battlespace

A systematic process of analyzing and visualizing the portions of the mission variables of threat/adversary, terrain, weather, and civil considerations in a specific area of interest and for a specific mission. By applying intelligence preparation of the battlefield/battlespace, commanders gain the information necessary to selectively apply and maximize operational effectiveness at critical points in time and space. Also called **IPB**. (FM 2-01.3)

intelligence requirement

1. Any subject, general or specific, upon which there is a need for the collection of information, or the production of intelligence. 2. A requirement for intelligence to fill a gap in the command's knowledge or understanding of the operational environment or threat forces. (JP 2-0)

Glossary

joint
> Connotes activities, operations, organizations, etc., in which elements of two or more Military Departments participate. (JP 1)

joint force
> A general term applied to a force composed of significant elements, assigned or attached, of two or more Military Departments operating under a single joint force commander. (JP 3-0)

joint intelligence preparation of the operational environment
> The analytical process used by joint intelligence organizations to produce intelligence estimates and other intelligence products in support of the joint force commander's decision-making process. It is a continuous process that includes defining the operational environment; describing the impact of the operational environment; evaluating the adversary; and determining adversary courses of action. Also called **JIPOE**. (JP 2-01.3)

joint operations
> A general term to describe military actions conducted by joint forces and those Service forces employed in specified command relationships with each other, which of themselves, do not establish joint forces. (JP 3-0)

measurement and signature intelligence
> Intelligence obtained by quantitative and qualitative analysis of data (metric, angle, spatial, wavelength, time dependence, modulation, plasma, and hydromagnetic) derived from specific technical sensors for the purpose of identifying any distinctive features associated with the emitter or sender, and to facilitate subsequent identification and/or measurement of the same. The detected feature may be either reflected or emitted. Also called **MASINT**. (JP 2-0)

open-source intelligence
> Information of potential intelligence value that is available to the general public. Also called **OSINT**. (JP 2-0)

operations security
> A process of identifying critical information and subsequently analyzing friendly actions attendant to military operations and other activities. Also called **OPSEC**. (JP 3-13.3)

personnel recovery
> The sum of military, diplomatic, and civil efforts to prepare for and execute the recovery and reintegration of isolated personnel. (JP 3-50)

signals intelligence
> 1. A category of intelligence comprising either individually or in combination all communications intelligence, electronic intelligence, and foreign instrumentation signals intelligence, however transmitted. 2. Intelligence derived from communications, electronic, and foreign instrumentation signals. Also called **SIGINT**. (JP 2-0)

technical intelligence
> Intelligence derived from the collection, processing, analysis, and exploitation of data and information pertaining to foreign equipment and materiel for the purposes of preventing technological surprise, asssessing foreign scientific and technical capabilities, and developing countermeasures designed to neutralize an adversary's technological advantages. Also called **TECHINT**. (JP 2-0)

Unified Action Partners
> Those military forces, governmental and nongovernmental organizations, and elements of the private sector with whom Army forces plan, coordinate, synchronize, and integrate during the conduct of operations. (ADRP 3-0)

References

REQUIRED PUBLICATIONS

These documents must be available to intended users of this publication.

JP 1-02. *Department of Defense Dictionary of Military and Associated Terms.* 8 November 2010.

JP 2-0. *Joint Intelligence.* 22 June 2007.

JP 2-01. *Joint and National Intelligence Support to Military Operations.* 5 January 2012.

JP 2-01.3. *Joint Intelligence Preparation of the Operational Environment.* 16 June 2009.

JP 2-03. *Geospatial Intelligence Support to Joint Operations.* 21 October 2012.

ADP 2-0. *Intelligence.* 31 August 2012.

ADRP 2-0. *Intelligence.* 31 August 2012.

RELATED PUBLICATIONS

These documents contain relevant supplemental information.

JOINT PUBLICATIONS

Most joint publications are available online:

<http://www.dtic.mil/doctrine/new_pubs/jointpub.htm>.

CJCSI 5221.01C. *Delegation of Authority to Commanders of Combatant Commands to Disclose Classified Military Information to Foreign Governments and International Organizations.* 10 September 2010.

JP 1. *Doctrine for the Armed Forces of the United States.* 20 March 2009.

JP 3-0. *Joint Operations.* 11 August 2011.

JP 3-05.1. *Joint Special Operations Task Force Operations.* 26 April 2007.

JP 3-08. *Interorganizational Coordination During Joint Operations.* 24 June 2011.

JP 3-13. *Information Operations.* 13 February 2006.

JP 3-13.1. *Electronic Warfare.* 25 January 2007.

JP 3-13.2. *Military Information Support Operations.* 7 January 2010.

JP 3-13.3. *Operations Security.* 4 January 2012.

JP 3-50. *Personnel Recovery.* 20 December 2011.

JP 3-60. *Joint Targeting.* 13 April 2007.

JP 5-0. *Joint Operation Planning.* 11 August 2011.

ARMY PUBLICATIONS

Most Army doctrinal publications are available online:

<https://armypubs.us.army.mil/doctrine/Active_FM.html>.

ADP 3-05. *Special Operations.* 31 August 2012.

ADRP 3-05. *Special Operations.* 31 August 2012.

AR 27-60. *Intellectual Property.* 1 June 1993.

AR 190-8. *Enemy Prisoners of War, Retained Personnel, Civilian Internees and Other Detainees.* 1 October 1997.

AR 380-5. *Department of the Army Information Security Program.* 29 September 2000.

AR 380-10. *Foreign Disclosure and Contacts With Foreign Representatives.* 22 June 2005.

References

AR 381-10. *U.S. Army Intelligence Activities.* 3 May 2007.
CALL Handbook No. 04-07. *(FOUO) Interpreter Operations Handbook.* March 2004.
FM 2-01.3. *Intelligence Preparation of the Battlefield/Battlespace.* 15 October 2009.
FM 3-05. *Army Special Operations Forces.* 1 December 2010.
FM 3-19.4. *Military Police Leaders' Handbook.* 4 March 2002.
FM 3-53. *Military Information Support Operations.* 4 January 2013.
FM 3-57. *Civil Affairs Operations.* 31 October 2011.
FM 27-10. *The Law of Land Warfare.* 18 July 1956.

OTHER PUBLICATIONS

DD Form 2745. *Enemy Prisoner of War [EPW] Capture Tag.*
DODD 5100.20. *National Security Agency/Central Security Service (NSA/CSS).* 26 January 2010.
Executive Order 12333. *United States Intelligence Activities.* 4 December 1981.
Intelligence Community Directive 301. *National Open Source Enterprise.* 11 July 2006.

REFERENCED FORMS

DA Forms are available on the Army Publishing Directorate Web site: <www.apd.army.mil>.
DA Form 2028. *Recommended Changes to Publications and Blank Forms.*
DA Form 4137. *Evidence/Property Custody Document.*
DA Form 4237-R. *Detainee Personnel Record.*

Index

A
all-source intelligence, 1-7, 2-16, 3-11, 4-3, 6-2, 7-3, 7-4, 7-7, 7-8, 7-13, 7-19, 8-3, 8-4, E-1

C
counterintelligence (CI), 1-2 through 1-5, 1-7, 2-5, 3-5, 3-6, 3-9, 3-10, 3-12, 4-3, 5-2, 5-3, 5-5, 6-2 through 6-4, 6-7, 7-2 through 7-6, 7-10 through 7-14, 7-17 through 7-19, 8-2, 8-5, 8-6, A-3, C-1, D-2, E-7

G
geospatial intelligence (GEOINT), 1-7, 2-4, 2-5, 2-14, 3-2, 3-3, 3-6, 3-10, 3-14, 4-3, 6-1, 6-4, 7-2, 7-5, 7-7, 7-12, 7-19,

H
human intelligence (HUMINT), 1-2 through 1-4, 1-7, 2-4, 2-5, 2-16, 3-5, 3-6, 3-10, 3-12, 3-13, 4-3, 5-3, 5-4, 5-7, 6-1, 6-3, 6-4, 7-3 through 7-7, 7-10 through 7-14, 7-17, 7-19, 7-21, 8-5, 8-6, A-3, A-4, C-3, E-1 through E-4, E-7, E-9, E-10

I
imagery intelligence (IMINT), 1-2, 1-10, 2-4, 2-16, 3-14, 4-3, 7-5, 7-9, 7-13, 7-14, 7-18, 7-19, 8-4, 8-6

intelligence disciplines, 1-1, 1-4, 1-6, 1-7, 2-15, 3-1, 3-6, 4-3, 5-2 through 5-4, 6-4, 7-15, 7-17, 7-19, 8-3, A-1

M
measurement and signature intelligence (MASINT), 1-7, 1-10, 2-4, 2-5, 2-16, 3-10, 3-13, 3-14, 6-1, 7-18

O
open-source intelligence (OSINT), 1-4, 1-7, 2-5, 2-16, 3-10, 4-3, 5-2, 5-3, 5-5, 7-17, 7-20, A-1, A-3, A-4, A-6

S
signals intelligence (SIGINT), 1-2, 1-4, 1-7, 1-10, 2-4, 2-5, 2-16, 3-4, 3-5, 3-10, 3-14, 4-3, 5-2, 5-5, 6-1 through 6-5, 6-8, 7-5 through 7-7, 7-11 through 7-13, 7-15, 7-16, 7-18, 8-3, 8-4, 8-6, A-3, E-6, E-7

T
technical intelligence (TECHINT), 1-2, 1-7, 2-4, 2-5, 3-12, 3-13, 4-3, 5-5, 7-20, E-6 through E-8

This page intentionally left blank.

ATP 3-05.20 (FM 3-05.120)
3 May 2013

By Order of the Secretary of the Army:

RAYMOND T. ODIERNO
General, United States Army
Chief of Staff

Official:

JOYCE E. MORROW
Administrative Assistant to the
Secretary of the Army
1300702

DISTRIBUTION:
Active Army, Army National Guard, and United States Army Reserve: Not to be distributed; electronic media only.

PIN: 103307-000